I0590518

The Best
AMERICAN
SCIENCE &
NATURE
WRITING
2025

The Best AMERICAN SCIENCE & NATURE WRITING™ 2025

Edited and with an Introduction
by Susan Orlean

Jaime Green, *Series Editor*

MARINER BOOKS
New York   Boston

FIRST EDITION

ISSN 1530-1508
ISBN 978-0-06-341421-1

Printed in the United States of America

25 26 27 28 29 LBC 5 4 3 2 1

Contents

Foreword   ix

Introduction   xiii

MICHAEL ADNO. *The Worm Charmers*   1
from *Oxford American*

ROSS ANDERSEN. *Do Animals Know That They Will Die?*   14
from *The Atlantic*

KATIE ENGELHART. *Letting Naomi Die*   21
from *The New York Times Magazine*

MARIA FARRELL AND ROBIN BERJON. *We Need to Rewild the Internet*   46
from *Noema*

RIVKA GALCHEN. *Pecking Order*   66
from *The New Yorker*

REBECCA GIGGS. *The Snake with the Emoji-Patterned Skin*   81
from *The New Yorker*

BEN GOLDFARB. *Relearning the Language of the Land*   98
from *Smithsonian*

ERIKA HAYASAKI. *Maui on Fire: When the Wildfires Came, a Young Couple Turned Toward Each Other to Survive Hawaii's Deadliest Natural Disaster*   113
from *New York*

FERRIS JABR. *The Whale Who Went AWOL*   132
   from *The New York Times Magazine*

DHRUV KHULLAR. *No Time to Die*   149
   from *The New Yorker*

SHARON LERNER. *Toxic Gaslighting: How 3M Executives
   Convinced a Scientist the Forever Chemicals She Found in
   Human Blood Were Safe*   162
   from *The New Yorker / ProPublica*

MAX G. LEVY. *I Tried to Train My Color Vision. Here's What
   Happened.*   184
   from *Sequencer*

EMMA MARRIS. *New York Is Wilder Than You Think*   192
   from *The New York Times*

JASON MAST. *This Father Built a Gene Therapy for His Son.
   Now Comes the Harder Part: Saving Others' Children, Too.*   196
   from *STAT*

TOM McALLISTER. *When We Used to Glow*   207
   from *Hippocampus*

DAVID NAIMON. *Eleven Stills*   210
   from *Prairie Schooner*

EMILY RABOTEAU. *Gutbucket*   216
   from *Orion*

SCOTT SAYARE. *The Smell Test*   229
   from *The New York Times Magazine*

LEATH TONINO. *This Is What It's Like to Camp in One
   of the Hottest Places on Earth*   245
   from *Outside*

SARAH ZHANG. *After the Miracle*   253
   from *The Atlantic*

Contributors' Notes   273

Other Notable Science and Nature Writing of 2024   277

Foreword

WHEN I WROTE my foreword to the 2020 edition of this anthology, I described the moment of writing—April of that year—as a strange time capsule. I wrote that the pandemic felt to be in full swing, but "this could just be the early ramp-up to something still unimaginable." Looking back, and remembering the panic I felt in those early weeks, I think the horror at least was close to peaking, even if the reality was not.

The unimaginable within which we find ourselves today, in the United States, is different. For starters, daily life, for many of us, is unimpacted. We go to work, we see friends, we grocery shop—we do not wipe down our groceries with bleach—and live our regular lives, just with the hum of alarm beneath it all. We do this knowing that for *other* people, in our country and elsewhere, normalcy is shattered. Maybe that's what people outside of New York City felt in April, but I was in the city then. This time, for now, I'm sheltered by my privilege. You may be too.

Furthermore, this moment is, unlike the pandemic, entirely precedented. A long legacy of authoritarian governments have left behind a playbook from which our government is working today. The deportations and abductions, the disinformation, the crackdowns on free speech. And, of course, the decimation of scientific research.

Governments fund scientific research because it benefits society. Not everything valuable turns a profit. (In fact, most valuable things don't.) And so, our erstwhile national leadership in science has depended on federal funding. Every scientist knows this, sustaining their labs from grant to grant. And if non-scientists didn't know this, hopefully they do now.

Science writing will not save us, but it is part of the fight. The pieces in this book were all written before Trump's second inauguration, making this volume a time capsule of its own. (Similarly, it was the 2021 anthology that captured that first year of Covid writing.) But even still, these pieces remind us of the value of science: for human progress, for the beauty of knowledge, and for our ongoing good work in the world.

Essays by Emma Marris and Ross Andersen remind us of our connections to other animals. Marris shows us that animals in cities are just as "wild" as their country cousins, in the process challenging us to rethink our boundaries between natural and human spaces. Andersen delves into the mystery of animals' awareness of death, challenging us to stretch our empathetic imaginings. Katie Engelhart similarly challenges us, in empathy and ethical possibility, with the harrowing complexity of the simple choice proposed in her piece's title: "Letting Naomi Die."

Three pieces also capture the limits and powers of scientific advances against diseases. In "The Smell Test," Scott Sayare captures the power and mystery of one super-smeller's discovery of a powerful application of her skill, with dramatic personal implications. Jason Mast writes of the lethal, heartbreaking tension between the need for specialized gene treatments and their lack of profitability in "This Father Built a Gene Therapy for His Son. Now Comes the Harder Part: Saving Others' Children, Too." In "After the Miracle," Sarah Zhang illuminates the rich complexities of a game-changing treatment for a disease that was once a slow death sentence.

And through it all, there is the fact that writing is a form of art. The most technical stories can be full of beauty, can evoke wonder. Essays like David Naimon's "Eleven Stills" and Tom McAllister's "When We Used to Glow" go further, using nature and science as fuel for empathetic engines. Emily Raboteau's "Gutbucket" interweaves meditations on her personal loss with a story of changing climate, evoking the powerful intermingling of grief and hope that is love.

Today, as I'm writing this, the cultural critic and historian (and my dear friend) Isaac Butler published an essay about the role of art in our political climate. He writes that he's noticed, lately, "a renewed feeling of pressure—mostly self-generated—to make work that Speaks To The Moment." In the midst of writing a history of the culture wars of the 1980s and '90s, he is finding that making such relevant art right now is actually painful and unrewarding.

Many science writers find themselves in a similar place. Sabrina Imbler, capable of great lyricism and joy in their writing about the world's weird creatures, has expanded their beat to profiling fired federal scientists. This is vital work. Their writing still sings. I haven't asked them how they feel about the work—galvanized or ground down or some combination of both—but, personally, when I see Blue Origin's space tourists doing PR for a company that wants to replace NASA's scientific mission with their profit-seeking, I wish my own writing about our uses of space could be less urgently relevant.

Butler writes in his essay, "[S]aving the world is not the artist's job. I've long been suspicious of people who talk as if it is. Art is the dream life of the self, and the collective unconscious of the culture. That is its job. And there's a lot of ways to do that job. Delighting is one. Comforting another. Challenging a third. Confronting a fourth. I mean there's many, many ways to do this. And many works of art do several at once."

Science writing is art. It is also information, argument, and activism. It kindles sparks of wonder and rings bells of alarm. It names and documents the horrors, and paves pathways toward hope. I don't know what the future brings, but I am so grateful for the writers in this book, and the many more beyond its pages, who are part of this legacy. I hope the future holds promise and justice for them, and for all of us.

JAIME GREEN

To submit work for consideration for future editions of this anthology, go to jaimegreen.net/basn.

Introduction

I DOODLED MY way through high school science, convinced that one-cell protozoa and gutted frogs had no place in my tender life. I considered myself a reader and a writer, a word-lover, and anything that wasn't on the page of a novel left me cold. In college, I was required to take a science class, and I barely squeaked by. Given my lack of scientific aptitude and appetite I wisely enrolled in the class so commonly known as "Rocks for Jocks" that I wonder if the University of Michigan eventually gave in and named it that in the official course offerings. The jocks were in the class because they couldn't be bothered with anything that required real academic effort. I was there because I was a liberal-arts romantic who believed science was dry and dense, so abstract and parochial that no one but a scientist would be animated by it or understand it. I figured I could at least manage to figure out a few chunks of stone.

Once I eked out a passing grade in Rocks, I thought I had put science behind me. Imagine my dismay when I began working on my book *The Orchid Thief.* I was convinced I could write about John Laroche's scheme to propagate rare orchids without appreciating—and then explaining to readers—the principles of orchid existence. But the minute I began reporting, I realized I had to refresh my measly grasp of botany. Can I confess to you that I was really nervous? As such a lackluster student of science, I was afraid I wouldn't comprehend it. Even worse, I wasn't sure I would be able to take what I might learn about orchids and write about it with the verve and swing I hoped to bring to the book. I'd have to try. To begin my tutorial, I found a few simple texts on the general subject of plant life, and a few about the strange and special world of orchids, and dove in.

When I finally raised my head, a month or two later, I was hooked. The insularity I'd always associated with science gave way to a new fascination. I will never fancy myself a scientist, but that research made me a science fangirl. The drama! The mystery! The near magic of it! I think I fell hardest when reading about Darwin's interest in a particular orchid with a column shaped so singularly—very long and very, very narrow—that no ordinary bird or bee could have pollinated it. Darwin posited that a moth or other insect with an equally long, extremely narrow proboscis that fit the orchid perfectly had to exist; otherwise, how could the species have thrived? In time, such a moth was discovered. Nature had made a matched set. To my eternal surprise, the sections of the book that delved deep into science were the ones I had the most pleasure writing. I loved taking the reader into the botanical world and sharing my delight with them; it was as if I'd learned a new language, and I wanted them to speak it too.

As it happens, I keep finding myself writing about science and nature in spite of myself. I wrote about dog breeding in *Rin Tin Tin*; I had to learn a lot about the science of fire for *The Library Book*. It has finally dawned on me that science and nature are not in a silo. They are a constant in our ability to understand the world, and they are part of many narratives where you wouldn't expect them.

Writing about science and nature for a wide audience is a special talent. I'm sure I'm not the only person who shrank from the subject as a young person or who approaches reading a science story worried it will be dull or daunting. The writer who takes this rich subject and writes about it learnedly but engagingly is performing a generous act, drawing in both the readers who are already game for the topic as well as the wary, the skeptical, the doodlers who can't remember anything from Biology 101. Among the most important missions of nonfiction writing is to open the reader to a new perspective on the world. Writing science and nature pieces is one of the highest forms of answering that mission, especially because many readers might begin those pieces worried that a nonexpert won't be able to penetrate a science-heavy story.

It's not only a generous act: It's an urgent one. Science and nature have always been important, but in the last few years, they have become vital. Certainly, since the pandemic, we've all become amateur scientists, whether we intended it or not. We have become fluent in spike formations of corona viruses and the hazards of wet markets and the dangers of wild weather and the peril of a changing climate. The exigencies of the moment have made us all freshly aware of how complex a system this world is, how many ways it can go wrong, how delicate the balance is. Covid may have been the first time many of us really understood the meaning of "public health" and "herd immunity." Climate change, which has been an abstract notion for quite some time, became manifest in recent years in concrete ways. We have had floods in seemingly unfloodable New York City, wildfires in watery Maui, collapsing cliffs in rock-solid Nantucket. Science and nature, always the backdrop of life on Earth, have roared into our everyday awareness, and it looks like they are going to stay.

There is beauty there too. Science keeps unlocking mysteries, revealing secrets, helping us heal. And as imperiled as nature seems, it remains amply, gloriously gorgeous: The world is still full of beauty. A few victories—a species saved, a wetland preserved—helps buoy us and remind us of how spectacular so much of the non-man-made world still is. It is nonfiction's task to point out nature's challenges, but also to remind us of why it deserves our care.

As a writer—and a reader—I'm partial to stories that do one of two things. The first is the story that delivers surprise, that introduces the reader to some aspect of life they never dreamed existed, or a person who is exceptional but unsung, or a subculture that has rarely been exposed. The writer serves as the intrepid explorer who has headed out into the unknown and returns to share the goods. Such stories abound in the worlds of science and nature, of course. There is so much newness, so much a reader who doesn't follow academic journals doesn't know, so many aspects of the natural world that are still being found.

I've chosen several such stories for this collection. Take, for instance, Michael Adno's wonderful piece, "The Worm Charmers," which appeared in the *Oxford American*. It tells the story of a Florida family who for three generations has perfected (and made a business out of) "worm grunting," a method of charming earthworms out of the mud. (The enchanted worms are then collected and sold as fishing bait.) Another marvelous example of this sort of piece (which also involves a long, skinny critter) is Rebecca Giggs's "The Snake with the Emoji-Patterned Skin," from *The New Yorker*. Did you know people collect ball pythons? I didn't. Did you know they go to great pains to breed "morphs"— pythons with designer skins that are polka-dotted or striped or covered in squiggles? Who knew? These are stories about nature, and about science, but most of all, they offer a chance to light up a new corner of the world; they expand our understanding of humanity.

The other type of story I love is one that takes a well-known subject and illuminates it in a new way. Stories about science and nature are especially tasked with this, since they often take topics with which the reader has at least a passing understanding. The best stories then deepen and expand on the subject. The magic of the piece is not that it presents something new, it's that it takes the familiar and finds a new way of thinking about it. In "New York Is Wilder Than You Think," Emma Marris takes a subject well-known to anyone who has spent any time in New York City—the presence of a whole range of animals and birds within the city—and contemplates it in a fresh way. What does it mean to be wild? What does it mean to be invasive? Are the pigeons and rats of the city less deserving than the hawks that show up now and then, or the coyotes that recently wandered down from Westchester? I had never considered many of the questions she raises in the piece, and it made me think about wildness, and in particular urban wildness, in a whole different way.

Science stories often concern themselves with the biggest, and also most intimate, subject of all: our bodies. The best stories about our flesh and bones, our mortal coil, are revelatory but relatable. Since Covid, we've all become more attuned to our physical selves, but how can we write about that in a compelling way that is instructive but also graceful? Many of the stories I've chosen to include here do that beautifully. Consider Dhruv Khullar's piece, "No Time to Die," about longevity evangelist Peter Attia, who argues that we can extend our lives—but at a considerable cost. Khullar taps into his own wish to live longer and be fitter in the piece, which raises it from a well-done profile into something much more ambitious, viewing Attia's attitude and recommendations within the context of our natural desire to live forever.

Another exceptional piece that centers itself on the human body is Erika Hayasaki's marvelous story for *MIT Technology Review* about rapid DNA analysis and its use in identifying the victims of Maui's wildfires. Hayasaki sets up the story as a mystery: A young man, Raven Imperial, who grew up in Maui but lives in California, is unable to reach any family members after the Lahaina fires to learn whether they survived. When the fires are finally contained, the death investigators arrive to identify remains. Until I read Hayasaki's piece, I had never fully understood how devastating it is for families to wait for those identifications, nor had I realized how long it takes—months or even years—using traditional methods. With the new methodology Hayasaki describes, which uses small machines that can be deployed in the field, identification can be done in as little as two hours. For mass casualty events—war, floods, fire, earthquakes—which are a dismayingly common occurrence, this kind of identification will be transformational. Even if they've been subjected to terrible forces, our bodies still keep our signatures, our stories, our identities. For people who have lost someone, this will ease a little of their pain.

Identity is at the heart of another one of the stories I've included here. The wonderful essay from *Sequencer* magazine by Max Levy describes the intricate, dizzyingly complex process involved in recognizing faces. As someone who is very good at recognizing faces but is married to someone who is horrible at it, I came to this piece with a stake. Is the ability to recognize and remember faces a sign of intelligence? Or is it just a fluke of how one brain is wired versus one that's set up differently? The delight of Levy's piece is that it reminds the reader how something that seems so ordinary—spotting people you know—is a highly choreographed firing of brain cells. It makes you appreciate that you are able to recognize anyone at all, and leaves you admiring what a good machine the human brain really is.

Science and nature writing sometimes calls for the human guinea pig approach: What could be a more direct way of reporting than to expose yourself to some condition, some experience, and then report back to those of us who would never dare to do the same? Madness, maybe, but the result can be a story that allows the readers to *almost* feel the experience for themselves. I sweated the entire time I was reading Leath Tonino's *Outside* magazine piece, "This Is What It's Like to Camp in One of the Hottest Places on Earth." I would not have arm-wrestled him for the assignment under any circumstances, but I'm very glad he undertook it. As a result, we now know, in searing detail, what it feels like to camp in 120-degree weather in the Mojave Preserve. Like the best stories, some of the pleasure in this piece is just reading great descriptions and well-paced narrative, but Tonino takes us almost viscerally somewhere we are unlikely to ever go. This is the marvel of reading, the boundlessness of the experience, the transitive nature of words on a page, thoughts being shared, stories being told.

Writing about science and nature is sometimes just a matter of literary handholding, walking an uninitiated reader into the thickets of research and theory. That's a worthy undertaking, but the best of it does so much more. Science isn't a partitioned subject; nature surely isn't a topic detached from the rest of life. They *are* life. They inform every aspect of how we live in the world. They are in the foreground or at least the background of every story we consider. Many of these writers might be surprised to find themselves called science writers. Most likely, they consider themselves writers on call for any story that catches their eye. The best of these pieces could just as easily be categorized in any number of other ways; the point is that they examine a topic and artfully deliver on it. They approach their subjects not as experts but as eager learners, delving into dense material and sharing what they've learned with us readers.

One of the stories here barely fits in the larger bucket of nonfiction. I'm thinking of David Naimon's piece, "Eleven Stills," which appeared in *Prairie Schooner*. Consisting of eleven short entries that very loosely consider trees, journeys, art, vultures, and the Tower of Silence, the piece could arguably be considered a prose poem or a meditation. It begins with an epigraph about air travel, taken from a journal on climate change, so you are tipped off at the start that the writer is considering the world and its physical systems. This is perhaps the most unusual of the pieces I've chosen here. Besides loving it as a piece of writing, I wanted to include it to push your understanding of what science and nature writing can be. It can be explanatory, but it can be poetic. It can be experiential, but also investigative. It can advocate, or it can just illuminate. It is expansive. Reading these stories made that last point exquisitely clear. Our understanding of our physical selves, and of the world and universe we inhabit, is the stuff of these stories, and so they are a wide-ranging bunch with a huge scope of interests. And so they should be. The best science writing—which these stories represent—reminds us of the wide world and our connection to it, and the multitude of ways we make our place in it.

Susan Orlean

The Best
AMERICAN
SCIENCE &
NATURE
WRITING
2025

MICHAEL ADNO

The Worm Charmers

FROM *Oxford American*

A HINT OF blue on the horizon meant morning was coming. And as they have for the past fifty-four years, Audrey and Gary Revell stepped out their screen door, walked down a ramp, and climbed into their pickup truck.

Passing a cup of coffee back and forth, they headed south into Tate's Hell—one corner of a vast wilderness in Florida's panhandle where the Apalachicola National Forest runs into the Gulf of Mexico. Soon, they turned off the road and onto a two-track that stretched into a silhouette of pine trees. Their brake lights disappeared into the forest, and after about thirty minutes, they parked the truck along the road just as daylight spilled through the trees. Gary took one last sip of coffee, grabbed a wooden stake and a heavy steel file, and walked off into the woods. Audrey slipped on a disposable glove, grabbed a bucket, and followed. Gary drove the wooden stake, known as a "stob," into the ground and began grinding it with the steel file. A guttural noise followed as the ground hummed. Pine needles shook, and the soil shivered. Soon, the ground glowed with pink earthworms. Audrey collected them one by one to sell as live bait to fishermen. What drew the worms to the surface seemed like sorcery. For decades, nobody could say

exactly why they came up, even the Revells who'd become synony-
mous with the tradition here. They call it worm grunting.

Audrey and Gary Revell took to each other in high school. In
1970 when Gary graduated, he asked Audrey to be his wife, and
they married at his grandfather's place down in Panacea, about
thirty miles south of Tallahassee. For his entire life, he'd lived on
an acre six miles west of Sopchoppy, Florida, in an area known as
Sanborn. The place is set deep in the heart of the Apalachicola
National Forest, a vast expanse of flatwoods and swamp that covers over half a million acres struck through with rivers. It's where
he and his siblings grew up in an old church building, where
his great-grandfather had settled after finding his way up Syfrett
Creek into the wilderness. It's where Audrey and Gary settled
after their wedding. "I was only sixteen, so I feel like I grew up
here," Audrey told me. Soon after, they started looking for ways
to make ends meet, and Gary suggested, "We might ought to
look into that worm thing."

His family was already deep into worm grunting. Three generations preceded him, and by 1970, his uncles Nolan, Clarence,
and Willie weren't only harvesting the worms to sell as bait but
were working as brokers with their own shops that distributed
the critters throughout the South. It didn't hurt that Audrey fell
in love with it immediately. The work was seasonal, busiest in
spring. During other parts of the year, their family trapped for
a living, dug oysters, logged, raised livestock, and set the table
with what they grew in their yard or caught in the water or in the
forest. "That's how we learned the woods," Gary said. "We went
in every creek, water hole, pig trail. You name it."

By the 1970s, the cottage industry had reached its peak. Then
Charles Kuralt arrived in 1972 to film a segment for his epony-
mous CBS show, *On the Road with Charles Kuralt.* The attention
led the Internal Revenue Service and the U.S. Department of
Agriculture to start regulating the harvest of worms, investi-
gating unreported income, and implementing permit requirements. Back then, the sound produced by grunters in the first
hours of daylight was as common as birdsong in this forest, and
hundreds of thousands of worms were carried out in cans. Folks
who once turned to grunting to make ends meet seasonally were

soon in the woods year-round during that decade, competing to summon the bait to the surface and sell to brokers among the counties set between the capital city and the Apalachicola River. Millions of worms left those counties bound for fishing hooks across America. Money followed the pink fever, but as with any rush, the demand eventually dimmed as commercial worm farms caught on and soft, plastic lures became popular.

By that point, Audrey and Gary had decided to shape their own outfit. His uncles had told them, *You ought to just think about keeping all that money to yourself.* The couple had grown tired of depending on others for work. So, they set up their own shop full-time, cultivated clients as far away as Savannah, and delivered bait all over the South, driving it themselves, or sending it north in sixteen-ounce, baby-blue containers via Greyhound buses. "All the money was coming our way, what little we made," said Gary. "We struggled with it for a long time, because when you get off the grid like that and try to do it for yourself and you're young, it's hard."

I wanted to know what spending their life in the woods hunting for worms meant, but I also wanted to know where this mysterious, artful tradition came from. In the UK, there are a handful of worm-charming competitions and festivals in Devon, Cornwall, and Willaston that began in the 1980s and another in Canada that started in 2012. I'd heard of similar events in east Texas, of people using pitchforks and spades as well as burying one stick in the ground and rubbing it with another to coax worms up to the surface. Later, I even found a newspaper clipping from 1970 reporting on the first International Worm Fiddling Championship, in Florida. I searched for a deep well of literature on the practice but found nothing. Certainly, worm grunting predated the Revells. But why did rubbing a stick stuck in the ground with a metal file conjure earthworms? The only way to understand was to follow the Revells into the woods.

In February, I carved out toward the Revells' place from St. Teresa, a strip of homes along the Gulf Coast. Going first through Tate's Hell, then turning west through the tiny town of Sopchoppy, I slipped into the forest as the distance between each home grew wider and wider. I found myself in a sea of slash and longleaf pine. Six miles later, I met Gary Revell in his driveway beneath an eastern redbud throwing its first spray of pink flowers. "Morning, Mike," he said with a contagious warmth. In their kitchen, I met Audrey, who had already poured a cup of coffee, set out milk and creamers, and had a jar of sugar in hand. A few minutes later, we piled into their truck and drove down a narrow vein of road near Smith Creek. A horned owl drew a line through the trees, where the yellow flowers of Carolina jessamine crawled over palmettos. Black water pooled in ditches alongside the narrow road lined with bald cypress and the periodic sweet bay magnolia. By the time we reached where we were going, I had no sense of how far we'd gone or where we were.

Although the northern borders of the Apalachicola National Forest press right up against the Tallahassee airport, the place is remote. Across nearly six hundred thousand acres, you could spend lifetimes trying to map its dizzyingly vast flatwoods, hydric hammocks, and cypress stands. Two hundred and fifty million years ago when our contemporary continents formed, Florida's peninsula broke off a fault line belonging to what's now West Africa; they share the same basement rock today. Fifty-six million years ago, as sea levels receded, the Suwanee Current flowed from the Gulf of Mexico across what's now Florida's panhandle, bisecting Georgia before running into the Atlantic. And over the next twenty million years, Florida appeared first as an island separated from North America by a sequence of patch reefs before sea levels continued to fall and a bridge formed with Georgia, revealing this very forest. A few thousand years later, the bones of the southern Appalachians, ground into dust by glacial erosion, washed out of the Apalachicola River Valley and formed barrier islands that rim Apalachee Bay today. That river carried sediment down through Georgia and into the Gulf, which flanks the western edge of the forest. And as you move east, the New River, the Ochlockonee River, and the Sopchoppy River flow through the forest made up of two districts. An archipelago of sinkholes and hardwoods is lacerated by thin roads that mirror oxbows in the rivers. In 1936, when the land was declared a national forest, it became one of America's southernmost pockets of wilderness and among the world's most unique ecosystems. As the Revells told me, many are afraid of the place, scared to step foot out of the car. "I've walked all over these woods, so I love them," Audrey said. "A lot of times when we'll be going to work in the mornings, we won't meet a single car. It's just nice being out here mostly alone. You know?"

That first morning I spent with them, the Revells made their way to a part of the woods called Twin Pole. The forest service had recently burned a block of woods there, which meant the ground would be clear and easier to work. As we got closer, I could smell the sweet fragrance of smoldering slash pines and palmettos. For centuries, pine scrub and prairie throughout the South has burned naturally and been torched deliberately, first by Indigenous peoples like the Timucuan or Apalachee and then later by ranchers and land managers to replenish the soil and promote growth. Worm grunters follow the forest service's burns like a compass, as the open ground makes it easier to spot worms and avoid venomous snakes.

"All right, Mama," Gary said to Audrey before changing into a pair of boots, fastening knee pads, and slipping on gloves. We walked through the burnt palmettos, coated in a film of black soot, before he pointed to a few holes in the soil. They were clues to where worms were and where they were headed. He took his stob, one his son had hewn out of black gum, and knocked it a foot into the earth with his steel file before rubbing the file against the stob's head. He called each pass a "roop." With every roop, he mirrored the sound himself, groaning first in a low pitch then ascending to an abrupt stop. Gary would roop, pause, tell a story, then start again. It didn't take long before a dozen large earthworms began crawling around the earth between us as Audrey gathered them by hand.

"Gary can call up any kind of animal," Audrey said. Screech owls, ducks, even a bull they once came across in the woods. Once, after he called to a quail, Audrey swears the bird landed on his head. I looked down as Audrey picked up worms and could see this was a corollary. As Gary rooped and talked, Audrey drew concentric circles around him, picking up the largest worms and carefully placing them in a one-gallon paint can. Audrey noted

the difference between worms—"milky" that are lighter in color
and frail, and dark pink worms that last longer on the shelf. Gary
roops most of the time, but Audrey does sometimes too. "They're
coming up tail first," Gary said. He gazed down and read the
ground: Here were some castings left by worms; some mounds
of fresh earth; a transition in the ground that meant prime mois-
ture. The Revells' intuition was like that of the fishermen they
were collecting bait for, a catalog of knowledge assembled from
spending time out here and bound together by deep curiosity.
Gary knocked his stob down against the serpentine root of a pal-
metto and demonstrated how to change the pitch. "When I see
that," he said, pointing to some larger holes, "I know he's right
here somewhere close."

With a couple of paint cans filled, about five hundred worms
in each, Audrey and Gary headed back to their truck, collecting
scraps of trash and some firewood along the way. An hour later,
they dumped their catch out in a shed where they store their
worms, counting them out by hand and then placing them in
five-gallon buckets filled halfway up with sawdust they collect in
the forest. Folks that know them come and collect worms from
the shed themselves, leaving the money they owe in a box on the
wall. Often, they'll leave notes scrawled on pieces of cardboard,
check registers, and even a cast-off piece of packing tape that
read, "I got 200. I paid back the ten I owe."

For two convenience stores in Wakulla County, Audrey and
Gary are the source for worms. At home, they pack the bait in
clear plastic cups with baby-blue caps and deliver them each
week. In the decades since the Revells struck out on their own,
the market has winnowed with the advent of artificial baits and
farmed nightcrawlers, and so have the venues to sell worms. In
good years, they earned thirty thousand dollars, according to a
2009 piece in the *Tampa Bay Times*, but they told me they didn't
want to discuss what they make today. Some years, they harvested
oysters for part of the winter and then baited throughout the
warmer months. The two found their way through, together, even
when bad weather, drought, and competition reshaped the way
they worked. They started traveling farther into Liberty County,
hiking deep into the flatwoods to avoid previously worked pieces

of land. In summer, when the temperature turned mean, they worked Tate's Hell at night. "This earthworm deal is something that you got to live with and stay on top of to be able to survive it," Gary said, "and we can say we've lived a very good life." They'd raised their two sons this way, spent their lives living with the forest, watching almost every sunrise out there together. "It ain't been no easy deal, but there's really nothing on earth I'd trade for it," Gary said. Today, one of the Revells' sons, who is now forty-eight, marks the fifth generation of their family collecting the pink currency from the forest.

In the nineteenth century, Gary's great-great-grandfather paddled up the Ochlockonee and into a branch that bent into the trees before it dissolved into a shallow stream. Audrey and Gary live in that area today, near a creek named for one side of his mother's family, the Syfretts. As kids, Gary and his two brothers, Lucious and Donald, came up in the woods, often passing the days with three cousins opposite the creek from them. "We didn't have a lot of people around, but we had this forest, and that kept us occupied." Their father, Frank, was an equipment operator for the county during the week, but worked alongside his brothers on the weekends, grunting in the forest at first light. Fifty years ago, he could earn as much as a hundred dollars in two days of baiting, which dwarfed what he made in a week for the county, roughly eight hundred dollars in today's money. Gary tagged along any chance he got. That's how he first heard the tale of his great-grandfather's worm discovery in the 1940s. Living along the Ochlockonee River, his great-grandfather fished often, and developed a sense of what baits worked where and when. While repairing his car one day, he'd left it running, jacked up the chassis, and removed a wheel. As the tire rolled away and his eyes followed it, he saw the ground strewn with pink worms.

As the story goes, his great-grandfather tested the theory elsewhere, leaving the car to idle and seeing worms sprout up on the spot. It was clear the vibrations stirred the worms, making it easier to collect bait and therefore sell it. This is how the mysterious practice became central to the Revells' lives.

Later, the men noticed worms appearing when they chopped wood or ran saws against saplings. Gary remembered using an axe handle as a stob, rubbing the blade of another axe against it. Some folks in north Florida called it worm fiddling, worm rubbing, worm snoring, worm charming, and, of course, worm grunting. Styles and materials for coaxing worms to the surface varied. Some people preferred hickory stobs and used steel leaf springs from cars as a file. The Revells used different-shaped stobs for different sorts of soil, but they always used black gum, persimmon, or cherry wood, and preferred flat, thick steel files.

What's strange is that despite the widespread practice of worm grunting, I couldn't find a definitive origin story. There wasn't a deep well of folklore to draw from online: not in the University of Florida's special collections archive, the Florida State University archives, or those of Florida Agricultural and Mechanical University. I searched my copy of the Federal Writers' Project's guide to Florida, organized by Stetson Kennedy and partially written by Zora Neale Hurston, with no luck. I couldn't find anything that went farther back than the 1970s. But after another pass through the newspapers at the University of Florida, I found a path that stretched back more than a century.

On Friday, July 16, 1946, the *Bradford County* (FL) *Telegraph* ran a front-page item, "Know Anything about 'Worm Grunting'?" They asked readers to submit letters, offering a five-dollar prize for "the best replies to a series of questions on this fascinating subject." Among them: how long had the practice existed, who told them about it, where they grunted, what they looked for, what they used, and what time was best to do it. Three months later, the paper published six letters. Dave Crawford from Starke wrote that he'd learned of it in 1933. Some claimed that it had existed at least since 1896, another since 1866, while one reader claimed it had been around in some form since 1786. One man wrote, "When I was a small boy, there was an old colored woman that worked for us. In the afternoon she would take me out and teach me to grunt for worms. She told me her mother taught her to grunt worms." Those anecdotal accounts raised the question of whether this was a tradition that extended back to the period of chattel slavery in America or even farther, before Indigenous peoples were forced from the land that settlers would come to call Florida.

The Revells' tales of grunting echoed those long-ago anecdotes. Readers referenced an axe handle method, or crosscut
saws, and an iron and a stake—all before Audrey or Gary were
born. The winning letter from Dave Crawford revealed a bit of
poetry and intuition that grunters still practice today: "When the
wen is from the west the werms come up good and when you see
the birds feeding on the ground and the red heads flying from
tree to tree you can grunt up better. Just get a old ax or tire iron
and a good pine stob about 2 feet long and a old lard bucket and
get down by the swamp where it is wet and boy go to rubing and
get busy and grunt long and loud and the old boys will come out
they hiding place."

That tradition endures, largely unchanged here in the Apalachicola National Forest. Yet, it's vanishing like so many other foodways, forms of heritage, and ways to earn a living in this part of the
country. Lots of folks preferred this work to other forms of labor,
such as driving an Uber in town or food delivery, but commercial
fishing, crabbing, and the shrimp industry have shrunk with each
passing year due to increasing regulation, depleted fisheries, climate change, and cheaper imported seafood. The same is true for
oyster harvesting, once a mainstay of the region's foodways. After
years of oyster decline partly due to overharvesting and negligent
water management, in 2020 the Florida Fish and Wildlife Conservation Commission mandated a five-year halt in harvesting oysters
from the Apalachicola Bay. It was part of a $20-million plan to re-
store the habitat and population. That ban promised to leave local
oyster tongers without work until 2025. As for worm grunting and
its slow decline, the passage of time is responsible too. "All the old
people is gone," Gary said. "That was the key to the whole thing.
They set it up."

In 2002, a committee was organized to preserve the tradition
and put on the first annual Sopchoppy Worm Gruntin' Festival.
Every second Saturday in April since, Rose Street and Winthrap
Avenue fill with vendors, bands, and demonstrations. There's a
ball and an annual queen. Media outlets flock to Wakulla County
to cover the festival, often centering the Revells in their pieces.
In 2009, they appeared on the Discovery Channel's *Dirty Jobs*.
That same year, Jeff Klinkenberg profiled the Revells for a cover

story in what is now the *Tampa Bay Times*. Nobody could say definitively why the worms responded to vibrations, though, until a neuroscientist arrived in Sopchoppy with a theory.

As a kid in Maryland during the 1970s, Kenneth Catania had a curiosity about the woods near his home that shaped his career path as a neuroscientist with a bent toward ecology and biology. His obsession with moles came later during a job at the National Zoo in Washington, D.C. And that obsession eventually grew into a dissertation on star-nosed moles, which revealed how their sensory cortex evolved and developed to process information. This, by proxy, revealed how all mammals' senses evolved. In 2006, he earned a MacArthur Fellowship or "Genius Grant." The award came with $500,000. Two years later, he headed for the Apalachicola National Forest, thinking that the moles there might help him unravel another mystery about a different group of underground creatures.

For years, he'd wanted to visit the worm festival in north Florida, but annual field work always overlapped. Finally, in 2008, he drove to meet the Revells in Sopchoppy. He arrived with a question shaped by a few sentences written a century earlier by Charles Darwin about worm behavior as it related to moles.

Darwin published his last book in 1881, *The Formation of Vegetable Mould Through the Action of Worms with Observations on Their Habits*. A sentence that struck Catania read, "It has often been said that if the ground is beaten or otherwise made to tremble, worms believe that they are pursued by a mole and leave their burrows." Darwin continued, "Nevertheless, worms do not invariably leave their burrows when the ground is made to tremble, as I know from having beaten it with a spade, but perhaps it was beaten too violently." Seventy years after Darwin's shovel experiment failed, Dutch biologist and Nobel Laureate Nikolaas Tinbergen claimed that herring gulls tapped their feet to drum up worms, employing "exploitative mimicry." By 1982, evolutionary biologist Richard Dawkins had built off that notion, staking claim to the idea of "rare enemy effect," by which predators cast themselves in the role of another predator to exploit their prey's behavior.

Then in 1986, a paper by John H. Kaufmann of the University of Florida drew a connection between wood turtles' stomping to draw worms to the surface and the work of worm grunters. "Many humans collect earthworms for fish bait by hammering or scraping on a stake driven into the soil. . . . There is now evidence that wood turtles, *Clemmys insculpta,* use the same principle in obtaining earthworms for food," Kaufmann wrote. He also noted an earlier paper from 1960 by Tinbergen that identified a corollary in herring gulls among other birds like flamingos and geese that drummed up prey by "paddling." Especially fascinating is that Tinbergen hypothesized that the worms mistook the birds' paddling for the vibrations of a mole. "That's what drew me down there," Catania told me. He wondered whether worm grunters were unintentionally mimicking a predator, possibly a mole like Darwin and Tinbergen suggested. "Nobody had formally studied it," he said.

On that first morning in Florida, Catania's alarm woke him at five. He got ready and met the Revells, who charmed Catania immediately as he took a seat in the cab of their truck. As they drove into the forest, he thought of this Darwinian theory that shaped his own hypothesis: that earthworms had developed an escape response to vibrations caused by a foraging mole. "What's beautiful about the system there is the earthworms are native, so they evolved there, and if the moles are there, they evolved there, too," Catania said. Most importantly, he wanted to find out if the vibrations generated by worm grunting echo that of a digging mole and, if so, how the earthworms respond.

As they rode along, Catania noticed mole tunnels crisscrossing the backroads. He saw more around the stand of trees where Audrey and Gary worked. Catania was spellbound as he watched the couple work. Weeks later, he returned with recording equipment, marking flags, and a garden trowel. He spent hour after hour, day after day, in the forest, dropping geophones into tunnel routes, hoping to record the vibrations of moles digging, as well as those produced by Gary's grunting. For every worm Audrey picked up, he placed an orange flag in the ground, mapping just how many worms appeared, in what directions, and how far from Gary's stob. Then, he stalked moles underground, using stakes placed along their routes to reveal where they were headed, and used the garden trowel to catch them. Back at the Revells' place, they took a handful of worms, placed them in a five-gallon bucket, and dressed them in a pile of sawdust. Catania picked up a mole and dropped it into the bucket. The worms fled to the surface. "Okay," Catania thought, "things are pretty clear."

He replicated this experiment in larger bins with controlled variables. The result was the same. As soon as the mole entered the soil, the worms fled to the surface. Catania later recorded the sound of an eastern American mole digging and compared it to his recordings of Gary rooping. It was a sonic match. The vibrations were almost identical.

Catania's work with the Revells confirmed Darwin's theory set forth more than 125 years earlier. Worm grunters had unknowingly applied "exploitative mimicry" like that employed by herring gulls or wood turtles to lure the worms to the surface. Catania published his paper that same year in *PLoS ONE*, a peer-reviewed journal. *The New York Times* even ran a small story about his findings, as did NBC News and other outlets. Before he returned to Nashville, Catania received a parting gift from Audrey and Gary—a rooping iron that had been in their family for decades. As he drove north that day, he stopped one last time in the woods, drove a stob into the soil, and rooped with a clear sense of what was happening underground.

On my final morning with Audrey and Gary, a seam of blue sky between the pines grew brighter as they drove out into the forest. Slowly, the first signs of light threw deep shades of purple against the clouds before pink, then scarlet bands passed through the trees. "That's beautiful," said Audrey.

They parked their truck along the road, collected their gear, and walked into the woods. As we neared a brake of trees, Gary passed me the stob and file, pointing to a patch of earth, and I clumsily drove the stob down. I tried to place my hands on the file the same as Gary, and I slowly slid the steel at an angle. A deep noise followed, and I just smiled, rooping again and again. I varied speed and angles, making some wince-worthy goose noises on bad passes, but I found a rhythm, and soon I'd drawn up a dozen worms. I moved a few times, continuing to work, removing some layers. When I finally got up, Gary asked, "So, Mike, what do you think?" My chest throbbed and sweat ran down my neck. "It's fucking hard work," I said.

Back at their place, Audrey made some sweet tea and showed me a couple albums of photographs she'd made of flora and fauna in the forest. She told me of terrestrial orchids "as pretty as one you would buy," of the pitcher plants in spring, and the white "worm flowers" that signal damp ground. "You never know what you might see," she said. Finally, she brought out some scrapbooks and clippings of articles from *The New York Times, Scientific American,* and the *Tallahassee Democrat.* In 2010, the Revells received Florida's Folk Heritage Award, an honor recognizing Floridians who preserve living traditions. Governor Charlie Crist presented the award in a ceremony at the state capitol. As we looked through those reminders of their life in the forest, Audrey and Gary turned serious. "I'm a steward of this forest," he said. "I don't do nothing to try to abuse it or change it." I asked Audrey what the forest meant to her. "Everything," she said.

That afternoon, as I prepared to leave, I found myself moved in a way I hadn't been in years, fascinated by their connection to the forest, aboveground and below. "As much as we've done it, I've thought, 'Man, you've got to be crazy,'" Gary said of their work. "But, if you take me away from it, I ain't worth nothing. I'm one of the last." I drove away with a sore palm and a cup of worms beside me.

ROSS ANDERSEN

Do Animals Know That They Will Die?

FROM *The Atlantic*

MONI THE CHIMPANZEE was still new to the Dutch zoo when she lost her baby. The keepers hadn't even known that she was pregnant. Neither did Zoë Goldsborough, a graduate student who had spent months jotting down every social interaction that occurred among the chimps, from nine to five, four days a week, for a study on jealousy. One chilly midwinter morning, Goldsborough found Moni sitting by herself on a high tree stump in the center of her enclosure, cradling something in her arms. That she was by herself was not surprising: Moni had been struggling to get along with the zoo's fourteen other chimps. But when Goldsborough edged closer, she knew that something was wrong. Moni had a newborn, and it wasn't moving.

Goldsborough raced downstairs to a room where the zookeepers were preparing food for the chimps, and told them what she'd seen. At first, they didn't believe her. They said that Moni was probably just playing with some straw. After the keepers saw the baby with their own eyes, they entered the enclosure and tried to take it away from her. Moni wouldn't part with it. They decided to wait and try again.

By this point, another female chimp named Tushi was lingering nearby. Tushi was one of Goldsborough's favorites. A few years earlier, she'd achieved global fame for executing a planned attack on a drone that was recording the chimps for a documentary. Long before that, she'd had a miscarriage of her own. For Tushi, the sight of Moni and her baby may have brought back that memory, or even just its emotional contours. For the next two days, she stayed near Moni, who held the tiny carcass. Finally, in a tussle with the keepers, it fell from Moni's grasp and Tushi snatched it up and refused to give it back. Moni grew extremely agitated. The keepers separated Tushi in a private room. Moni pounded at the door.

Goldsborough wasn't sure how to interpret this behavior. Moni seemed to have been driven by fierce maternal attachment, an emotion that is familiar to humans. Tushi could have been responding to an echo of this feeling from deep in her past. But it's not clear that either of the chimps really understood what had happened to the baby. They may have mistakenly believed that it would come back to life. It's telling that we can't say for certain, even though chimpanzees are among our nearest—and most closely watched—neighbors on the tree of life.

This past June, more than twenty scientists met at Kyoto University for the largest-ever conference on comparative thanatology—the study of how animals experience death. The discipline is small, but its literature dates back to Aristotle. In 350 BCE, he wrote about a pair of dolphins that he'd seen gliding beneath the surface of the Aegean Sea, supporting a dead calf, "trying out of compassion to prevent its being devoured." Most of the literature in comparative thanatology consists of anecdotes like these. Some are short, like Aristotle's, but others, like the story of Moni and her baby, which was published in the journal *Primates* in 2019, and to which we shall return, contain extraordinary social details.

Scientists would like to go beyond these isolated scenes. They want to understand what feelings surge inside animals when they lose kin. They want to know whether animals are haunted by death, as we are. But they're hampered by certain practicalities. They cannot interview animals (or at least not yet). They can monitor their hormonal shifts—baboon cortisol levels spike when they lose someone close—but these can be triggered by other stressors. They don't give us the texture and grain of their grief, if indeed it is grief that they feel.

So far, the best comparative-thanatology data has come from observations of animals in the wild or captive populations in zoos. But here, too, there are problems. The species that react most interestingly to death—the usual suspects: nonhuman primates, whales, and elephants—have long lifespans. Their communities don't lose individuals very often. Capturing systematic data about their reactions to death tends to require years' or decades' worth of work.

Alecia Carter, an evolutionary anthropologist at the University College of London, told me that she has identified a colony of more than a thousand rhesus macaques on Cayo Santiago, an island off Puerto Rico, that would be perfect for such a study. The monkeys are highly social, and tend to live for fifteen or twenty years—long enough to form deep relationships, but not so long that their deaths would be too few and far between. As a start, one of Carter's grad students recently spent nearly a summer there collecting data. Only eleven monkeys died. "It was a great season for them, but terrible for us," Carter said.

Humans have spent months in steamy jungles or zoo enclosures, dodging feces, to pursue this work. We are death-obsessed animals, after all, and have been since the dawn of recorded history, if not before. Our oldest work of epic literature tells the story of King Gilgamesh and his struggle with mortality. "Death is sitting in my bedroom, and wherever I turn, there too is death," he says, before setting out in search of a plant that promises immortality. Human cultures have devised richly symbolic rituals to precede death and to follow it. For more than ten thousand years, we have laid our lost children in the ground, surrounded by flowers. We are a species of faithful mausoleum attendants, pyramid builders, inventors of the three-volley salute. We have imagined a great many afterlives for our dead, in heaven above or here on Earth aboard the great turning wheel of reincarnation. We have sicced our philosophers, armed with fine distinctions and caveats, on death; their definition of it now runs to more than ten thousand words. We have even projected our finitude onto the universe itself. Scientists tell us that it too will die after the last galaxies unwind and the black holes evaporate, particle by particle, trillions upon trillions of years from now.

These elaborate human conceptions of death are not passed down through our genes. They develop over decades in the minds of individuals, and in our cultures, they accrete over centuries. Human children tend to learn that death is not a temporary or reversible state somewhere between the ages of four and seven, or a bit earlier if they lose a beloved family member or animal. A 2004 paper in *Cognition* argued that, at this developmental stage, children understand death as a permanent loss of agency.

In her new book, *Playing Possum: How Animals Understand Death*, the Spanish philosopher Susana Monsó argues that many other animals likely share this simple concept of death. That may seem like common sense, but without access to their minds, it is difficult to know for sure. Mammals, fish, birds, reptiles, and insects are all cognizant of agency in the natural world. They monitor their environments for movement. They distinguish between inanimate objects and those that crawl or swim in pursuit of some goal. And some of them behave in ways that suggest an understanding that other animals can lose this agency forever. The hard part is knowing whether these behaviors flow from a conceptual recognition of death, or if they're simply instincts.

Consider the termite. At the June meeting in Kyoto, an urban entomologist at LSU named Qian Sun presented a paper on the corpse-management practices of the eastern subterranean variety. More than a million of these insects may pack into labyrinthine underground colonies that sprawl for hundreds of feet. When worker termites come across a dead colleague in one of the colony's tunnels, they react in different ways, depending on the state of the corpse. Fresh ones, they devour. Old and moldering ones, they bury. Other social insects that live in close quarters engage in similar practices. (Aristotle noted that bees carry their dead out of the hive.) But these behaviors don't appear to be driven by a concept of death. Termite corpses produce oleic acid, which appears to trigger the burial behavior, as it does in several different social insects. When E. O. Wilson dabbed this chemical onto a live ant, its fellow colony members did not pause to consider whether the still-moving animal had suffered a permanent loss of agency. They simply carried it outside, even as it kicked its legs in protest.

Chimps are not termites. Their large, complex brains are better-equipped to entertain a concept like death, and there is evidence to suggest that they feel something like grief. Several species of nonhuman primates have been known to gather around a community member that has recently died. In many cases, they will touch its body gently. These gatherings tend to dissipate slowly and in a patterned way: the individuals who were closest to the deceased animal stay longest. Jane Goodall observed an eight-year-old chimp lingering by his dead mother so long that he died too.

Other mammals also tend to congregate around their dead. When giraffes do it, they swing their long necks at scavengers to keep them at bay. In India, the bodies of five young elephants have been found with branches and dirt scattered over them, leading some scientists to suggest that they'd been buried. André Gonçalves, an expert in comparative thanatology from Kyoto University, cautioned me about making too much of this anecdote. The elephants were found in trenches that they may have fallen into, he said. The dirt and branches could have piled up as family members tried desperately to dig them out.

In her book, Monsó argues that too much has been made of all these grief responses. She reminds her readers that animals live in a bloody world where predators pounce in the dark of night, or plunge down, talons-first, from unseen heights. The lurid violence of their environment provides a rich text for understanding death. Monsó imagines a young stag watching a dominance struggle between two older bucks. After their horns crack together a few times too many, the weaker combatant fails to get up. The young stag begins to understand the basics of mortality. If the lesson doesn't take, he will likely have many occasions to relearn it.

This education would presumably be accelerated in carnivores, who see death frequently and at close range. Gonçalves told me that he's not so sure. Many animals eat other animals while they are still alive, he said. It's not clear that they are trying to bring about death, or that they conceive of it as a separate state of being. They might simply be trying to get a moving food source into their mouths, like frogs that shoot their sticky tongues at everything moth-like, just as a matter of reflex. Gonçalves noted that even the precise one-bite kills deployed by big cats are instinctual, not learned behaviors.

Among chimpanzees, acts of wanton violence, up to and including murder, suggest a deeper understanding of death. Like wolves and lions—and people—chimps sometimes team up to kill members of rival groups. These attacks can have an air of premeditation. Two or three males will cross into terrain occupied by another group. They will move quickly and with stealth, and won't stop to eat, even when passing by prime food sources. They target lone victims, and coordinate their attacks to avoid sustaining bruises or cuts of their own. In some cases, they will keep on striking long after a victim has signaled submission and let up only when the unlucky animal has ceased to breathe.

If indeed chimpanzees do have a concept of death, it is not as layered or intricate as ours; that much is certain. Humans know what death is, and we know that someday it will happen to us. James Anderson, an emeritus professor at Kyoto University, who is widely regarded as the godfather of comparative thanatology, has argued that chimps do not have a similar sense of their own mortality. He does not believe that anyone has ever really seen a chimp attempt suicide, in all the many thousands of hours that we have observed them. According to Anderson, only an animal that knows that it can die will try to bring about its own death. That there are no reliable reports of chimps, he says, or any other animals, engaging in this behavior suggests that the existential burden of mortality is uniquely ours to carry.

Anderson doesn't know for sure, of course. Comparative thanatologists aren't really in the business of giving answers, at least not yet. They can tell us that a chimp's conception of death is grander than a termite's, but much else is mysterious and maybe always will be. We can only hope that by continuing to watch chimps, we will notice new behaviors that betray a bit more of their interiority, or at least give us new grounds to speculate. The story of Moni and her baby may be one of them. Before coming across it, I'd read many papers about the way that chimps react to their deceased, but very few about how they treat the bereaved.

After the zookeepers got Tushi alone, they decided to let things cool off. They kept her away from the others until the next day. In the meantime, for Moni, everything had changed. She had previously struggled to connect with her fellow chimps in the enclosure. She had a way of pulling other females' fur too hard during grooming, and she often sat too close to them, staring awkwardly. On the day that Tushi rejoined the group, Moni was surrounded by the other chimps. When she saw Tushi, she leapt up to perform an aggressive threat display. She even slapped her.

Tushi didn't fight back, and in the thirty days that followed, she and the other chimps interacted with Moni more than they ever had before. No other chimp experienced an equivalent increase in attention. Almost all of the chimps contributed. They embraced Moni and gave her extra body kisses. But they did not contribute equally. Some cared for Moni more than others, and none more than Tushi. Something important seems to have passed between the two chimps. A few months later, things largely went back to normal in the enclosure. Moni stopped getting extra kisses. The males started bullying her again. But she and Tushi still often sat together. Even today, I am told, they remain close.

KATIE ENGELHART

Letting Naomi Die

FROM *The New York Times Magazine*

THE DOCTORS TOLD Naomi that she could not leave the hospital. She was lying in a narrow bed at Denver Health Medical Center. Someone said something about a judge and a court order. Someone used the phrase "gravely disabled." Naomi did not think she was gravely disabled. Still, she decided not to fight it. She could deny that she was mentally incompetent—but this would probably just be taken as proof of her mental incompetence. Of her lack of insight. She would, instead, "succumb to it."

It was early 2018. She had come to the hospital voluntarily, because she was getting so thin. In the days before, she had felt her electrolyte levels dip toward the danger zone—and she had decided that, even after everything, she did not want to be dead. By then, Naomi was thirty-seven and had been starving herself for twenty-six years, and she was exquisitely attuned to her body's corrupted chemistry. At the hospital, she was admitted to the ACUTE Center for Eating Disorders & Severe Malnutrition for medical stabilization. There, doctors began what was once called refeeding but is now more commonly called nutritional rehabilitation, using an intravenous line that fed into her neck. Reintroducing food to an emaciated body can be dangerous and even lethal if done too quickly. Physicians identified this phenomenon in the aftermath of World War II, when they observed skeletal concentration-camp survivors and longtime prisoners of war eat high-caloric foods and then drop dead of cardiac failure.

"Well, here I am," Naomi said in a video message that she recorded for her parents. "I am alive, but am I happy? I don't know. . . . It's pretty pathetic. I don't know how I feel about the fact that I would have died had I not come." In the video, she was wearing a hot-pink tank top, even though it was cool in the hospital room, because she wanted to shiver, because shivering burned calories.

A few days later, when she was not imminently dying anymore, Naomi announced that she was going home—and the hospital responded by placing her on a seventy-two-hour mental-health hold. Clinicians then obtained what Colorado calls a short-term certification, which required, by judicial order, that Naomi be detained and treated, in her case until she reached what physicians determined to be 80 percent of her "ideal body weight." In Colorado, as in most states, a patient can be treated against her will if she is mentally ill and found incapable of making informed decisions. That day, Naomi was transferred to a residential program at Denver's Eating Recovery Center (ERC).

"I'm so mad, I'm so mad," Naomi said in another video message, her voice dull and impassive. "I was completely disrespected. I was tricked." Naomi could feel that her mind was diminished—it was too slow, too slack—but she found that she could think in a straight line. She could reason. So why did the doctors claim otherwise? By then, she had been in and out of hospitals and psychiatric wards and eating-disorder programs, including the ERC, more times than she could recall. Was it really so irrational for her to assume that trying the same treatment for the hundredth time would be futile?

When she was a teenager, Naomi believed that treatment programs might save her. She ate supervised meals and attended group-therapy sessions where, among other things, patients discussed the origins and possible psychic functions of their eating disorders. Sometimes Naomi told the story of how she stopped eating because she thought it would make her a faster swimmer.

Or the one about how she just wanted to be special, like her eldest brother was special because he was so smart. Other times, she told the story about the day her grandfather died and the whole family went to eat at a restaurant. Naomi was revolted watching everyone nourish their bodies with something as carnal as food when they should have been awash in grief. Years later, it was hard to tell if any of these origin stories mattered. With each inpatient admission, Naomi gained weight. Each time, the extra weight felt unbearable, and she lost it soon after discharge.

As the years passed, Naomi found it harder to be "compliant" with standard treatment. She refused to participate in group sessions. Or she disengaged during therapy, which she found infantile and pointless. She sometimes tampered with her intravenous lines, because it was too awful to watch those plastic bags of liquid calories empty into her body. During some admissions, Naomi forced herself to gain weight so that she could be discharged. Other times, she signed herself out against medical advice. Later, Naomi started bingeing and purging. She would excuse herself after meals and step into the backyard to vomit into plastic bags that she would throw into the neighbor's yard, so that nobody would see. She vomited and vomited until stomach acid burned through the enamel of her teeth and she had to spend $22,000 to replace them.

In between treatment programs and emergency hospitalizations, Naomi, at eighteen, went to college. She wanted to study psychology, but all she could really do was exercise for hours a day after eating almost nothing: maybe an apple. In her final year, she dropped out. Later she found jobs that she cared about—a certified nursing assistant who did home health assessments, a patient coordinator at a hospital—but they were often interrupted by yet another medical admission.

As she moved through adulthood, Naomi acquired new diagnoses: anorexia binge-purge type, osteoporosis, hypotension, gastroparesis, superior mesenteric artery syndrome, obsessive-compulsive disorder, post-traumatic stress disorder, bipolar disorder. She took mood stabilizers and antidepressants and anti-psychotics. Her bipolar manic periods felt like an ecstatic embrace of the world. The depressed periods made her want to kill herself, and sometimes try to.

She collapsed into her thirties. She had no hobbies and no friends. She had become a kind of professional patient: her whole life whittled down to the airless world of her diseases, the logistical management of her self-denial. Everything was epic drama, but also staggeringly boring. To Naomi, her doomed attempts to get well had started to feel less tragic and more ridiculous. It wasn't so much that she wanted to be dead, at least most of the time. It was that she could no longer stand anyone trying to cure her—especially because the "cures" were always the same and never worked. "I'll either die of anorexia or I'll die of suicide," Naomi told me when we first spoke. "I've accepted that."

After her admission to the Eating Recovery Center, Naomi spent a few days lying in bed, being fed by a nasogastric tube, which pushed fluids and nutrients down her throat and into her stomach. Some days, she put plastic flowers in her hair and took selfies, just frowning at the camera. She made conversation with her roommate, who was very nice but sometimes threw up on the floor between their beds. After a few weeks, Naomi gained enough weight that she could be discharged into an outpatient program. It was there, she says, that a therapist asked her if she had ever heard of palliative care.

The field of palliative care was developed in the 1960s and '70s, as a way to minister to dying cancer patients. Palliative care offered "comfort measures," like symptom management and spiritual guidance, as opposed to curative treatment, for people who were in pain and would never get better. Later, the field expanded beyond oncology and end-of-life care—to reach patients with serious medical illnesses like heart disease, HIV and AIDS, kidney failure, ALS and dementia. Some people who receive palliative care are still fighting their diseases; in these cases, the treatment works to mitigate their suffering. Other patients are actively dying or in hospice care. These patients are made "comfortable," or as comfortable as possible, until the end.

Naomi's therapist had printed out an article for her to read. It was called "Medical Futility and Psychiatry: Palliative Care and

Hospice Care as a Last Resort in the Treatment of Refractory Anorexia Nervosa," published in 2010 in *The International Journal of Eating Disorders.* The paper's authors argued that psychiatry needed its own subfield of palliative care: specifically for the 15 to 20 percent of patients whose anorexia developed a "chronic course" and did not respond to standard treatment—and for the fraction of those patients who did not want to keep trying to get better.

These patients, the paper proposed, should not be coerced into treatment but offered an approach that aimed to palliate their psychological pain—until, maybe, they died of their eating disorders. The authors acknowledged that the idea of letting a mentally ill person decide to withdraw from treatment was uncomfortable, even radical—even though the rest of medicine already recognized a patient's right to stop fighting her disease and risk dying. A patient with advanced kidney failure, for instance, might become exhausted and decide to quit dialysis treatments. "It has been argued that patients with anorexia nervosa should have similar rights to discontinue treatment, despite the fact that in their case food refusal might seem irrational. Although patients with anorexia nervosa may irrationally choose not to eat, they are often competent to make decisions in all other areas of their lives."

When Naomi looked up the paper's authors, she was surprised to find that one of them, Dr. Joel Yager, was based in Denver. He was a psychiatrist at UCHealth University of Colorado Hospital and had started working with anorexia nervosa patients in the 1970s. Back then, psychiatrists were just beginning to understand anorexia as a mental illness, one with neurological and metabolic components. Nevertheless, there was reason to be optimistic; with early and aggressive treatment, a vast majority of the starving young patients got better.

Of course, there were the ones who didn't. Within the treatment community, anorexia had always been described as an acute condition, something with an adolescent onset and relatively short duration. It was only in the mid-1980s that a small number of academic articles began to refer to a "protracted" or "long-term course" of the disorder, and then eventually to "severe and enduring" anorexia. It was this kind of patient, typically a woman with a decade of failed treatments behind her—"kind of hobbling along in life," Yager said—who found her way to him.

Yet when Yager, who was then working at the University of California, Los Angeles, looked for guidance on what to do for such a person, he found almost nothing. All he could see were articles instructing him on how to exert his will over recalcitrant patients, how to give them more standard treatment aimed at full weight restoration. And sometimes, because that was all he had to offer, his patients would simply stop coming to appointments. Yager would discover, later, that they had gone home and died alone on their sofas. Maybe by starvation, maybe by suicide. Maybe in pain. "I felt like a failure," Yager told me. "They fired me, basically, at the end, knowing that I wasn't able to help them anymore and wasn't eager to just see them through the end." In a desperate attempt to not abandon them, he had abandoned them. Bludgeoned them with care. Rescued them to death.

He came to think that he had been impelled by a kind of professional hubris—a hubris particular to psychiatrists, who never seemed to acknowledge that some patients just could not get better. That psychiatry had actual therapeutic limits. Yager wanted to find a different path. In academic journals, he came across a small body of literature, mostly theoretical, on the idea of palliative psychiatry. The approach offered a way for him to be with patients without trying to make them better: to not abandon the people who couldn't seem to be fixed. "I developed this phrase of 'compassionate witnessing,'" he told me. "That's what priests did. That's what physicians did 150 years ago when they didn't have any tools. They would just sit at the bedside and be with somebody."

Yager believed that a certain kind of patient—maybe 1 or 2 percent of them—would benefit from entirely letting go of standard recovery-oriented care. Yager would want to know that such a patient had insight into her condition and her options. He would want to know that she had been in treatment in the past, not just once but several times. Still, he would not require her to have tried anything and everything before he brought her into palliative care. Even a very mentally ill person, he thought, was allowed to have ideas about what she could and could not tolerate.

If the patient had a comorbidity, like depression, Yager would want to know that it was being treated. Maybe, for some patients, treating their depression would be enough to let them keep fighting. But he wouldn't insist that a person be depression-free before she left standard treatment. Not all depression can be cured, and many people are depressed and make decisions for themselves every day. It would be Yager's job to tease out whether what the patient said she wanted was what she authentically desired, or was instead an expression of pathological despair. Or more: a suicidal yearning. Or something different: a cry for help. That was always part of the job: to root around for authenticity in the morass of a disease.

Most of the patients who asked for palliative care, Yager thought, probably wouldn't want to die but would be open to dying if it meant that they could stop trying to get better in the same old ways. Yager imagined that his practice would, in large part, be defined by absence. No coercive care. No obligatory weekly weigh-ins. No heroic measures. A palliative approach might even mean de-prescribing drugs that helped keep a mental illness at bay but made the patient feel bad in other ways: prioritizing comfort over life extension or symptom reduction. The care would be shaped by what the patient wanted, in the moment.

From Denver, Yager started publishing papers about his ideas, and other doctors started contacting him, clinicians who had, in the quiet context of their own practices, invented a kind of palliative psychiatry of their own. Once in a while, Yager heard directly from a patient.

"Dear Dr. Yager," Naomi wrote in an email in February 2018. "After 20 years of trying the same thing over and over again and expecting different results, I am tired of fighting the system."

After he read Naomi's email, Yager called her. "Come in," he said. "Let's see." With her tangle of disorders, Naomi presented as a complex patient—but only in the way that many other patients were complex. She was depressed and bipolar, but both conditions were being managed with drugs. Naomi told Yager that her current outpatient providers would continue treating her only if she strove for and ultimately maintained 80 percent of her ideal body weight—but that she couldn't meet their condition because she couldn't bear to be so heavy. "I've been there, I've done that," Yager remembers her saying. "I have these obsessions. They won't let go of me. Nothing they have ever given me in therapy has ever changed those internal, infernal thoughts."

Yager agreed to help Naomi put together a palliative-care team at UCHealth and to oversee her psychiatric care. It was obvious that, in many ways, Naomi's thinking was deeply distorted—but when she expressed her desire to stop fighting, Yager thought she seemed "as clear as a bell."

Contrary to what medicine had recognized for most of its history, Yager knew that a substantial number of patients with psychiatric disorders were, in fact, medically and legally capable of making decisions on their own. When given a standard "capacity test"—which measures a patient's ability to understand information related to a specific decision, appreciate benefits and harms, reason, and express a choice—many passed. In one study of seventy adult women with severe anorexia, forty-six were found to have "full mental capacity."

If a patient is found capable, her physician is meant to respect her choice, whether or not it seems rational or circumspect. The test is always whether a person *is able to reason*, not whether she seems reasonable to her doctors.

After their initial meeting, Naomi was told that she could set the rules. Point 1 was no more residential programs, ever. "It only accelerates the suffering," she said. "And I refuse to encounter it ever again." Point 2 was no involuntary heroic measures from her doctors, no mandatory weigh-ins, no behavioral therapy. Naomi was willing to play around with new psychiatric medications—because, she said, a better drug might make her remaining days more tolerable—but she no longer wanted to analyze the root causes of anything. She was tired of telling her life story, tired of trying to interpret things.

Naomi told her new palliative-care physician, Jonathan Treem, that she could not increase her weight, at least not without something bad happening. She believed that whenever she relaxed a bit on the anorexia front, her bipolar disorder got worse; whenever she gained a few pounds, it threw her mood way off-kilter—

and that was worse than starving. She needed to appease both demons.

Naomi was willing to accept the odd temporary measure, like an infusion of electrolytes to lift her energy, but she wouldn't treat her underlying physical disorders: her osteoporosis or her gastrointestinal issues or whatever else set in. Fixing those things would do nothing for her mood. Besides, at some point her body would fail and it would be inevitable, and she would let it happen. "If my heart decides that it's done beating," she said, "then I will not stop it."

When Treem sat with Naomi, he could feel "an incredible agony that was internalized and unremitting and, to a certain degree, barely endurable"—a depression that was "likely perennial and unlikely to be subject to change." In Treem's view, Naomi's anorexia was both a cause of pain and a symptom of a larger hurt. "She's actually used her body as a communication tool for a long time. 'I want to look so grotesque that people cannot look away.'"

Treem was an internal-medicine doctor by training, and most of his work involved palliating patients who were dying of typical somatic ailments: cancer, heart failure. Working with Naomi, he found, required him to undertake some "philosophical groundwork." He thought about how he might protect his patient from her most self-destructive impulses, but also refrain from bulldozing over what she wanted. Treem talked with Naomi about how choosing to die from the natural progression of a disease was not the same thing as suicide.

To Treem, it felt as if Naomi was asking for something more than his nonintervention; she wanted his mercy. His permission to let go, his compassion. It made him think about the other doctors who had treated her. "This is where it gets into a passionate discussion," he told me. "If you are going to accept responsibility for the people you save, and you're going to elevate them as examples of why everyone should undergo compulsory treatment, you had better recognize the blood on your hands. That, on some level, in order to 'save everyone,' you are perpetuating suffering in others."

Yet Treem had his limits. He told Naomi that he could not look away if she was actively suicidal. Several times, after an especially unsettling appointment, Treem walked her down to the emergency room, where she was put on a seventy-two-hour mental-health hold.

Naomi also met regularly with Yager, who sometimes wondered whether, paradoxically, giving up recovery-focused treatment could steer his patient back to health. Palliative care, Yager reasoned, might give Naomi the cognitive space to reset. It would eliminate the classic power struggle between flailing eating-disorder patient and exacting psychiatrist and, perhaps, let her sense of fight turn inward. But Yager knew he had to be restrained in this thinking. If he approached Naomi's palliative care as a means to a cure, then it wasn't really palliative care at all—just a stealthy treatment program. This required a sort of intellectual sleight of hand. Yager had to be equally accepting of either outcome: that Naomi lived or that she didn't.

Besides, what did the alternative look like? Would he be better off to declare Naomi incompetent? Sedate her? Restrain her physically or chemically? Get court orders for involuntary medications and involuntary tube feeds—which wouldn't "cure" her anyway but would keep her alive for more treatment? Lock her on a ward? Try to keep her there? Hope she comes around? "Are you going to do it forever?" Yager asked.

Yager had always been suspicious of psychiatry's affinity for hope, of the hopefulness that many doctors deliberately exhibited for their patients. "I'm full of hope," he told me. "I'm one of the most hopeful guys you're going to find. But I'm also a realist."

Many psychiatrists, Yager knew, believed that they must hold hope for their hopeless patients, that a projection of hope, by a clinician, mattered—that it was even essential—because the hope could be absorbed by a patient and, in turn, change the course or constitution of her disease. In this way, psychiatry was fundamentally different from other kinds of medicine. In oncology, for instance, a doctor's professed hope for a patient could not shrink a tumor or lower a blood-cell count. But maybe, in psychiatry, there was a more porous boundary between physician and patient, between an illness and a patient's ideas about it. Maryrose Bauschka, a psychiatrist at the Eating Recovery Center, told me,

"I think there's often a lot of fear that if we're transmitting anything less than a message of hope—or anything less than, like, a full-court press—that we're not going to help them get better."

But couldn't a doctor's hope also be a kind of harm? Yager could see that some of his patients benefited from his cheerleading. Others, though, were propelled into unwanted treatment by somebody else's hope for them—and then left to feel defeated when it didn't work. So couldn't it also be argued that a doctor had a moral obligation not to provide hope that was unjustified, and maybe even to expose false hope where it lay? "We thus find ourselves in a paradox," wrote Justine Dembo, a psychiatrist and assistant professor at the University of Toronto, "in which hope is vital for recovery but may also lengthen lives of unbearable mental anguish. What is an ethical therapist to do?"

Yager knew that the evidence base for many recovery-oriented therapies—some of which had been in existence for decades—was weak. He had never found a single randomized control study proving, with any certainty, that the by-then-ubiquitous residential eating-disorder program worked better than other kinds of care. Many of the country's largest treatment facilities were owned by private companies that did not, as a practice, invite third-party researchers to study their approaches or track their long-term patient outcomes. Yager worried that the many doctors pushing residential programs were compromised, if not financially then at least intellectually. They had become, as he put it, "zealots for the model."

And there was certainly no evidence at all that a fourth, or fifth or tenth, attempt at the same kind of program was likely to be helpful, especially if the patient didn't want it. The same was true of involuntary care. There was some evidence that forced treatment could be life-sustaining in the short term, but its long-term effects were more uncertain. In his own academic articles, Yager wrote about the "willfully blind Pollyannish therapeutic attitudes" of psychiatrists throughout history, and of their "excessive hyperinterventionism."

Within the rest of medicine, "medical futility" had become a subject of contention in the 1980s, after relatively new interventions like cardiac life support and mechanical ventilation allowed

the nearly dead to be resuscitated and sustained. Sometimes, patients' families demanded that their loved ones be treated aggressively and kept alive, hearts beating and lungs pumping, when
there was no realistic prospect for recovery. Or alternatively, families pushed back against a physician's aggressive, almost knee-
jerk use of technology to sustain a flailing life. Eventually, those
doctors grew accustomed to admitting defeat, to acknowledging
that yet another week of life support or another round of chemotherapy or another aggressive surgery would serve no therapeutic purpose.

But the idea of futility remained "relatively unknown in the
world of psychiatry," according to a 2023 paper in *Frontiers in
Psychiatry*. When I asked a psychiatrist with expertise in severe
and persistent mental illness how much time had been devoted,
during her more than a decade of medical training and residency, to learning about futility, she laughed. "Zerooooo."

After all, in psychiatry, there were always more drugs and drug
combinations to try. More behavioral interventions and therapeutic modalities to employ. More clinicians who believed that
they alone had the special therapeutic touch. It seemed to Yager
that despite what every honest psychiatrist should know, psychiatrists were never really *allowed* to acknowledge futility—and so
never allowed to stop treating. In turn, their patients were never
"allowed" to say no. Never allowed to decline care. Certainly
never allowed to die.

In one 2023 study, published in *The American Journal of Bioethics Neuroscience*, 174 U.S. psychiatrists completed a survey on
"their attitudes about the management of suicidal ideation in
patients with severely treatment-refractory illness." The doctors
were given one of two case studies: the first, about a patient with
borderline personality disorder; the other, about a patient with
major depressive disorder. They were told that the patients had
already received every treatment that might reasonably be expected to work and that, despite this, they remained sick. The
psychiatrists were then asked to rate the expected helpfulness
of further treatments—and the likelihood that they, personally,
would prescribe them.

The conclusion of the study was stunning: "Sizable minorities of participants said they were likely to recommend interventions they thought were unhelpful." The authors identified several potential reasons. Perhaps the doctors were trying to meet expectations: the patient's, her family's, their colleagues', the system's. Perhaps they worried about legal liability.

But maybe there was another explanation. Maybe this was just the logic of a profession that saw death as the absolute worst outcome, regardless of what living might look like.

Physicians in the field had heard the emerging calls for palliative psychiatry, some with alarm. The idea that certain patients would be better off if they gave up on cure-focused treatment was, as Dr. Agnes Ayton of Britain's Royal College of Psychiatrists told me, "dangerous nonsense." For some of these doctors, Yager's writings about palliative psychiatry were not just ill-defined but threatening to the profession, particularly because they were so underdeveloped and so contentious and because, nevertheless, Yager and others were already deploying them.

Some physicians had doubts about the premise—core to Yager's thinking—that patients who were very sick could still have the mental capacity to make decisions as grave as the one to stop recovery-oriented care. A typical anorexic patient had cognitive distortions and pathological values. She was intransigent, fearful, cognitively inflexible. She could be emotionally anesthetized too, so apathetic that she didn't care very much what happened to her. Her brain was literally starving. How could such a patient be taken at her word when she said she was prepared to die—that it was what she "wanted"? Any experienced physician should know that what the anorexic patient "wanted" was perverted by her disease. He should see through the ruse—even if, like many people with anorexia, his patient spoke well and dressed well, was not in the depths of psychosis and could clearly articulate the potential medical benefits and drawbacks of various treatments. This was not mental lucidity, but instead a pantomime of reasoned thought.

Other psychiatrists took issue with the way Yager conceptualized futility. With anorexia nervosa, it was almost always impossible to say that a given treatment would be physiologically futile, because there was virtually no point at which an eating disorder became physically resistant to healing. If a patient ate, nearly all of her medical conditions could be reversed. It was even hard to make educated guesses based on how other patients had fared in similar situations, because there was so much variability between treatment programs and because nobody was collecting large databases of patient outcomes.

For the anorexic patient, any conclusions about "futility" would have to be based on fuzzier judgments about how a treatment might affect her quality of life. To critics, this was insufficiently rigorous. "Medical futility," the psychiatrist Cynthia Geppert warned in a 2019 handbook, "can only be tentatively and tenuously translated into psychological constructs."

In Yager's model, decisions about futility seemed to rest a lot on what the patient believed the effect of a treatment would be. But many people with chronic mental illness are ambivalent about recovery and resistant to treatment. They "know" that they will never get better. They "know" that a treatment will fail. These feelings are literally products of a pathology. This pathological despair must be challenged, not interpreted as an expression of enlightened thought and then honored in the name of patient rights.

"What many in the profession would say," Thomas Strouse, a psychiatrist and palliative-care physician at UCLA, explained, "is that anorexia leading to death is a form of protracted suicide." In this view (which Strouse does not endorse), accepting a patient's slow death by starvation and choosing not to medically intervene, with force if necessary, was akin to collaborating in a suicidal act. At the least, it was colluding with a person's mental illness.

Already, research showed that some patients with eating disorders who were involuntarily treated did well. In the short term, their rate of weight restoration was the same as that of voluntarily treated patients. One paper noted that among those admitted to the hospital, "nearly half of patients with eating disorders who denied a need for treatment on admission converted to acknowledging that they needed to be admitted within two weeks of hospitalization." The food, in other words, brought the insight.

Other physicians emphasized the current inadequacies in American mental-health care as a reason any futility judgment would be ethically tenuous. A decision that further treatment was "futile," they argued, would be meaningless if the patient had never received high-quality care in the past. In the case of eating disorders, many people can't access evidence-based treatment or experienced providers, because they don't have private insurance to cover it. Others do have insurance but discover that their providers' patience is limited. Patients are discharged from programs because their insurance companies do not believe that they are progressing quickly enough. Or because they seem to progress too quickly. These patients are released as soon as they have gained sufficient weight (as defined by the insurance company) but before their weight is fully restored. They then go home and get sick again. Can a person's decision to decline treatment, made in the context of resource scarcity, really be described as a free choice?

And the sickest of patients can still get better—even after *decades* of failed treatment. One study of adult patients with anorexia, published in *The Journal of Clinical Psychiatry* in 2017, found that nine years after the start of their illness, only 31.4 percent had recovered—but that by twenty-two years, the recovery rate had doubled to 62.8 percent. "These findings," the study's authors wrote, "should give patients and clinicians hope that recovery is possible, even after long-term illness, suggesting that even brief periods of weight restoration and symptom remission from anorexia nervosa are meaningful and may be the harbingers of more durable gains to be made ahead."

Angela Guarda, a professor of psychiatry and behavioral sciences at the Johns Hopkins School of Medicine, told me that palliative measures can sometimes be useful—but only alongside curative care and never instead of it. Guarda said she has treated several thousand patients with anorexia and still "cannot predict who will get better and who will not." Patients sometimes surprised her. So "how do I decide which patients of mine I should instill hope in, and which patients of mine I should decide to help die?"

In this way, critics argued, psychiatry was being mischaracterized by Yager's views. It wasn't that psychiatrists were bludgeoning chronically ill patients because they couldn't acknowledge their own shortcomings, or couldn't respect an anorexic person's wishes, or didn't have the empathetic imagination required to take pity on their patients. Doctors who refused to give up on treatment did not lack humility; they kept trying precisely because they had it.

Naomi's parents, Evelyn and Hal, first heard about palliative care in a hospital conference room, where different members of the medical team told them what they did and what was going to happen to their daughter. At first, Evelyn was just confused. She told Naomi that if something had a chance of working, then it was worth trying, "even though you might not want to do it, but come on. Let's try again." Maybe Naomi was just the unfortunate patient who took thirty years to figure out how to help herself.

The thing was, there were things Naomi hadn't tried. Her parents wondered if meditation or yoga might bring her some peace, but Naomi always said that none of that stuff worked for her. She didn't like to journal. She didn't believe in acupuncture. For a while, she took oral ketamine, which can have a rapid antidepressant effect in some patients, but it destabilized her moods. She was too afraid to try psilocybin. Hal, an engineer, often spoke of "coping mechanisms." He didn't think his daughter would ever be cured, but he thought she could develop "not just one mechanism, but maybe multiple mechanisms that she'd have in her little toolbox . . . so hopefully she could pull herself out of a tailspin."

"But that's never really been the case," Evelyn said. "She's never been able to pull herself out."

Benjamin, one of Naomi's brothers, mostly felt bad for his parents: two elderly people who were left alone to absorb Naomi's chaos when they were supposed to be living out their golden years. But he also wondered if his parents had enabled their daughter with all the emotional and financial cushioning that, in the end, had done nothing more than just barely keep her alive. At times, Benjamin urged his parents to seek legal guardianship over his sister so they could force her into treatment. But the whole process had seemed daunting.

In Colorado, a court-appointed guardian can have a patient forcibly tube-fed—but only through a special court order and only until the patient is medically stable and no longer at imminent risk of dying. And while guardians can order short-term measures, they can never compel a ward into long-term psychiatric care.

After starting palliative care in 2019, Naomi quit her job as patient coordinator at a hospital. She went on Medicare and Medicaid for disability and moved back into her parents' small house in the Denver suburbs. Naomi knew that her parents didn't accept the palliative approach. How could they? "They hold out hope," she said, "and they are hopeful that something will click in my head."

Naomi's room at her parents' house was tiny. It had three of Evelyn's quilts hanging on the walls, patchworks of black and white and red, and a TV by the doorway. There were piles of stuffed animals on the floor. There was no bed. Naomi preferred to sleep in a brown recliner, elevated, to keep the stomach acid down after purging.

Sometimes Naomi sat in the recliner and read through her electronic medical notes. Because she was on palliative care, Naomi's doctors were prompted to indicate, after appointments, whether they expected her to live longer than six months. Sometimes they did, and sometimes they didn't.

From there, she spent a year cycling in and out of the hospital, not eating for long enough that she might come close to dying and then agreeing to go to the hospital for emergency intravenous nutrition—but then, often, discharging herself after just a few days. Sometimes, during those admissions, Treem said, he was chastised by the attending physicians. They wanted to know why Naomi was at the hospital all the time. Why her nutrition wasn't being managed. "Why are we allowing her to continue to flounder?"

Then came the winter of 2022, which was a very bad winter. Within a six-week stretch, Naomi was hospitalized four times for suicide attempts. Then there were more. Sometimes someone would call her an ambulance. Other times, she would get scared and call one herself. The episodes involved different things: weed killer, benzos, batteries. "For some reason, I'm really obsessed with swallowing things," she said. Once, she poured bleach in her eyes. She didn't think it would kill her, but she liked the idea of going blind and not having to look at herself anymore. She did not go blind.

The attempts were never planned in advance, and later, Naomi could never quite reconstruct her thinking. In February, she ended up in the intensive-care unit on a ventilator. That winter, the bipolar disorder seemed to eclipse the anorexia. "I'm at the end stage with the bipolar," Naomi told me. "Like, I'm at the end. I mean, I'm just there." She had come to terms with the fact that she would probably die of starvation; now she thought she would kill herself instead.

Part of the problem was that the eating disorder had turned on her. At some point, Naomi had lost the ability to purge. She would try and try, but the vomit would not come. Her doctors explained that this happened sometimes with chronic patients. Unable to purge, a kind of mental pressure built up inside her. Her moods dipped.

Her parents weren't sure what to make of the suicide attempts. "They've been feeble," Hal told me. "In other words, she hasn't thrown herself off a thousand-foot cliff, jumped out of an airplane without a parachute." Then again, he said, "they're serious attempts." Some caused damage. Some didn't work but theoretically might have.

Yager agreed that, at times, it seemed as though Naomi was trying to "play Russian roulette" with her life. "There are some patients who are fatalistic and throw it up to God," he told me.

"That's the grayness," Treem said. "How do we know when Naomi has made a rational decision about being in line with her values versus when she is reacting to a torment in herself?"

At home, Naomi talked about "a decline in cognitive functioning." She spoke of being "flat" and "cloudy." She did not think that she was irrational, but as the terrible winter turned to spring, she had started to feel a little blurry.

What she really couldn't stand was how pointless it all felt. If there were at least a point to her suffering, then maybe she could bear it. But Naomi did not believe in salvation through struggle. She did not believe that her misery would lead her to some other, better place—or even to an enlightened understanding of things. She had no consoling story to tell about it. In fact, she had no story to tell at all, because a story needed a plot, and her entire life had just been the same awful thing.

*

In February 2022, Yager co-published a new paper titled "Terminal Anorexia Nervosa." The article, whose lead author was Jennifer Gaudiani, an internal-medicine physician who founded an out-patient eating-disorders clinic in Denver, proposed that psychiatry recognize a new clinical disorder, terminal anorexia, which would apply to the small fraction of patients for whom "recovery remains elusive" and palliative measures were not enough.

The paper offered several criteria for determining terminality: that a patient have a diagnosis of severe and enduring anorexia nervosa; that she be at least thirty years old; that she have previously attempted "high-quality" eating-disorder care; that she have consistent decision-making capacity and be prepared to die; and that she "understand further treatment to be futile." The label was important, the authors reasoned, because it would grant sick patients a formal diagnostic acknowledgment that they were dying, making it easier for them to access hospice care—and even, should they want it, and should they live in a state where it is legal for terminally ill patients, and should their physicians be willing, a physician-assisted death. (A year earlier, Canada revised its national medical-assistance-in-dying law, expanding the eligibility criteria to include people whose only condition is mental illness. The law will take effect in March.)

The paper presented three case studies: all deceased patients of Gaudiani's. One was a thirty-six-year-old woman named Jessica, whose eating disorder began in her junior year of high school, when she tried to lose weight for a vacation. Jessica had anorexia and obsessive-compulsive disorder, and she abused laxatives, sometimes taking a hundred tablets in a single day. In her twenties and thirties, she tried several treatment programs but usually left them early and against medical advice. She grew despairing and suicidal, once buying a gun and driving to a bridge where she contemplated jumping. "Fearful of suffering a long, drawn-out death from starvation," Jessica met with Gaudiani to ask for an assisted death. Gaudiani and a palliative-care physician agreed that forced treatment was likely to be futile and that Jessica would most likely die of the physical effects of her eating disorder within six months. Gaudiani signed the paperwork for

Jessica's death. Jessica took weeks to fill her prescription for lethal drugs and many more weeks to take them—lying in bed, holding her parents' hands. Even in the final month of her life, she forced herself to walk for hours a day to stay thin.

Yager and Gaudiani acknowledged that there were no explicit physiological markers of terminality in anorexia, no set point at which a patient could not possibly recover. For them, this fact did not preclude the possibility of a "terminal" diagnosis.

But for many readers, the paper had a warped logic. In its formulation, it was the patient's perspective of her illness, rather than the illness itself, that mattered. A mentally ill person was terminal, in large part, if she said she was.

Some of Yager's colleagues moved quickly to denounce the paper. Several journals published counterarticles: "Terminal Anorexia Is a Dangerous Justification for Aid in Dying," "Terminal Anorexia Nervosa Is a Dangerous Term." Everything that made Yager's model of palliative care alarming was in that paper—but made worse because the sick patients were bestowed with a medical label that validated their most deranged belief: that they were literally impossible to heal. Patricia Westmoreland, a Denver-based forensic psychiatrist who focuses on the ethical dilemmas around eating-disorder care, told me that the ideas in the paper were "absolutely unconscionable."

Most of the criticism focused on the case-study patients who were approved for assisted death. Everyone wanted to know how this could have possibly happened because, as of yet, there were no professional standards governing when, if ever, an anorexic person should qualify for assisted dying. And American psychiatrists had barely even broached the subject in theoretical terms—because, among other things, was it even legal?

In states with legal "medical aid in dying," a person had to be terminally ill and within six months of a natural death to qualify. But these patients' doctors had gone ahead—gone rogue—and proceeded anyway, after first inventing a medical term, "terminal anorexia," to cover their backs. Critics wondered what the follow-on effects would be: Would schizophrenic and depressed people eventually receive a doctor's help to die? Some noted that the case-study patients did not receive very thorough treatment before being declared "terminal." One had never completed a residential program.

Philip Mehler, the founder and medical director of ACUTE, the hospital unit where Naomi was admitted for medical stabilization in 2018, was one of the detractors, though he told me that he has "a lot of respect for Joel Yager," who is "a very thoughtful guy." For several years, Mehler has watched the debate about palliative psychiatry bleed out of academia and into the patient population. Patients with anorexia, he said, are often keen students of their disorder; they read the academic literature about it. As a result, for the first time in his career, he has had patients in their twenties ask about palliative care and assisted deaths. "So I think this has had a bit of a contagion effect," he said.

Other critics worried that the terminal diagnosis would exert a kind of downward pressure, that once labeled terminal, a patient would feel a certain obligation to become the thing she was described as being. Guarda, the Johns Hopkins professor, said she fears a future in which "patients actually become invested in acquiring the diagnosis." In which being terminal is an aspiration, and patients do whatever it takes to earn the title. Guarda can imagine those patients showing up at her office, waving a copy of the terminal-anorexia paper in hand. "See?"

Yager had little patience for the frenzy. He was particularly bothered by criticism that "terminal anorexia" was too hazy a definition and too unmathematical to be operationalized. That was literally every diagnosis in psychiatry. "Anorexia nervosa" was once a contested term too—there were long arguments about how its borders should be defined—and it had shifted over time. Even beyond psychiatry, medical diagnoses were always socially mediated in some way; every diagnostic label came from imperfect humans with imperfect evidence. Yager gave me the example of hypertension. "There are working groups of experts who are cardiovascular people, trying to define when it is that you have 'high blood pressure,'" he said. "The arguments are intense in R and D, because drug companies make money depending on how these criteria are defined."

In somatic medicine too, a patient's decisions and beliefs

could affect whether his condition was considered terminal. When a patient in renal failure decided to stop using dialysis machines, for instance, it was his choice that rendered him "dying" and "terminal."

What's more, in somatic medicine, a patient didn't need to have a good reason for stopping care. She didn't even have to try getting better in the first place. A cancer patient could decline chemotherapy that would very likely save her life. Because she didn't think the benefit was worth the pain. Because she wanted to go home to her children. Because she preferred to be treated by a homeopath. She could do what she wanted, just because she wanted to. Why should patients with mental illnesses be held to a different standard?

Yager was also frustrated by critiques that referred to larger problems with the mental-health-care system. Colleagues kept telling him that eating-disorder care wasn't good enough or accessible enough to allow for a terminal diagnosis—but what were they proposing in the meantime? That patients be made to suffer because the rest of us haven't done enough to help them yet? And anyway, an oncologist would never deny end-of-life care to a lung-cancer patient who wanted to stop chemotherapy, on the grounds that, say, the patient didn't previously have access to high-quality smoking-cessation programs.

Jennifer Gaudiani, Yager's co-author, told me that she has asked her critics directly: What would you have done differently with those patients? "And it can't be, 'I would change the eating-disorder system to be more inclusive and accessible.' Nope. We've got a patient in front of you, right now." Should she be abandoned in the name of ideological purity? Gaudiani believes that the paper's detractors demonstrate "an important flaw" in their logic: If a patient elects, willingly, to go into standard eating-disorder treatment, her decision is never scrutinized and her capacity is never questioned. But if, instead, the patient's decision is "incongruent with life-saving, then we question," she said. "That's not ethical."

"It doesn't make sense," Yager agreed. "They're 'incompetent' unless they want treatment?" His critics, he said, had no data at all to back up their claim of universal incapacity among anorexic people. Existing studies showed the opposite. Yager thought his critics were suffering from "positive outcome bias"; they remembered the patients who were saved and were grateful for it, but not the ones who died slowly and suffered all the while.

By late 2023, though, amid all the furor, even Gaudiani was walking back parts of the paper. Her own criteria for terminality, she told me, were "too inclusive"—and the phrase "terminal anorexia" was so controversial that it had, itself, become "harmful." (Gaudiani still believes that some patients with eating disorders should have access to physician-assisted death.) Yager told me that he does not regret what he wrote. "The main point is that some people die from the disease," he explained. "We have to be caring and attentive to them to the end."

Already, Yager could feel some of the paper's ideas burrowing their way into psychiatry. At the 2022 annual meeting of the American Psychiatric Association, there was a session on palliative psychiatry for severe and persistent mental illness. At the 2023 Royal College of Psychiatrists Conference in London and the Australia and New Zealand Academy for Eating Disorders gathering in Queensland, participants discussed, respectively, "the controversy on terminal anorexia nervosa" and palliative models for "end stage" eating disorders. That same year, the American Psychiatric Association's annual conference held a panel discussion titled "Physician Aid in Dying Based on a Mental Disorder."

In July, a professional gathering on palliative psychiatry was held in Toronto. Sarah Levitt, a psychiatrist at Toronto's University Health Network and one of the organizers, told me that the process of framing "palliative psychiatry" might pose a broader challenge to the profession. "There is maybe some interest in bringing conversations of death and dying into psychiatry," she said. "How might we do that?"

"I just feel like there's so many things going on. And all I can think about right now is the fact that, since I was last here, I've lost six pounds."

In September, four years after starting palliative treatment, Naomi was seated in a small hospital meeting room, across from Treem and a chaplain named Beth Patterson. The room was unadorned, apart from a single framed photograph of the Colorado mountains in winter. Treem and Patterson looked at Naomi, who looked down at her hands: red and cold from her low metabolic rate. She wore black shorts and had purple plastic flowers in her hair. When Naomi arrived at the hospital, an intake nurse asked her if she had felt down or depressed over the last few weeks, and Naomi said, "No," because whenever she answered "yes" she had to fill out a questionnaire.

She did feel depressed, though. The drugs for her bipolar disorder couldn't seem to lift her up, though they did flatten out the periods of mania. For several months, she hadn't been able to shower, because she couldn't stand to undress in the bathroom, which had mirrors in it. She was angry too. ("Bipolar rage," she said.) And tired. Naomi had started purging again—had, out of nowhere, regained the ability to purge. Slowly, she started losing weight. She wondered if she would lose more. In her case, she said, the bulimia looked nothing like the way it did in movies: some delicate woman excusing herself from the dinner table for a quick, discreet regurgitation. Naomi's episodes involved vomiting for hours until there was nothing but bile, and then vomiting that up too. Sometimes she saw blood in the toilet. "My esophagus is becoming what they call 'flappy,'" she told me. It was hard to swallow.

At the appointment, Naomi mostly wanted to talk about her parents: how there was something the matter with her dad's heart, how her mother was growing frailer. Naomi had started doing more around the house to help them. She cleaned and took care of the yard. She cooked them dinner—sometimes complicated recipes, from pictures she found online. She never ate any of what she cooked.

"This is a bit of a precipice," Treem said. "I don't know which way it moves." He wondered aloud whether the needs of Naomi's parents could compete, in some useful way, with her eating disorder. Could Naomi's transition from sick patient to caregiver be elevating for her? Could it be the resolution of her story—or maybe even the point of it? Or would the whole situation just break her?

"We have been looking for a long time for meaning in your suffering, right?" he said. "Your suffering exists. It's not going to change. It's a part of this disease, and the disease doesn't change. So the agony of suffering, the pain, is permanent. And living with that, without meaning or purpose to it, is demoralizing, corrosive. You get to this place where you're like, 'I can't do this anymore.' And you want to die and try to kill yourself, right? That's the dynamic. That's a part of this. So the question has always been: Is there a reason in this suffering? So that it can feel justified in some way."

"We have talked a lot about the crisis of dying," Treem continued. "We haven't really talked about the crisis of survival. How that might be really painful, and really difficult. Maybe even more so."

"Yes," Naomi said softly.

"Yes," the chaplain agreed, leaning in. "Yet here you are."

On some days, Naomi thought about assisted dying and whether she might qualify for it one day. It would be better to die that way than the other ways she had tried. An assisted death, at least, would be clean and painless. It would mean that someone had given her permission.

Naomi had read Yager's paper about terminal anorexia. She liked parts of it but thought it was incomplete. The paper proposed all these criteria for terminality, but it didn't include suffering. The assisted-dying law in Colorado didn't mention it either. "That's completely ridiculous," she told me. Naomi knew that she was no longer starved to the point of bodily collapse, like the case-study patients in the paper—but still she suffered terribly, and shouldn't that matter just as much? And really, shouldn't it matter as much as having an inoperable tumor or a failing heart? Shouldn't it matter the most?

"Let's presume you don't die," Treem said toward the end of the hourlong appointment. He asked Naomi if she could imagine looking back, ten years from now, and being able to say, "That was a good life."

Naomi looked so startled by the question that everyone laughed a little. But, no. She couldn't imagine that. "I can't," she said. "I can't imagine continuing on."

MARIA FARRELL AND ROBIN BERJON

We Need to Rewild the Internet

FROM *Noema*

The word for world is forest.
—Ursula K. Le Guin

IN THE LATE eighteenth century, officials in Prussia and Saxony began to rearrange their complex, diverse forests into straight rows of single-species trees. Forests had been sources of food, grazing, shelter, medicine, bedding, and more for the people who lived in and around them, but to the early modern state, they were simply a source of timber.

So-called scientific forestry was that century's growth hacking. It made timber yields easier to count, predict, and harvest, and meant owners no longer relied on skilled local foresters to manage forests. They were replaced with lower-skilled laborers following basic algorithmic instructions to keep the monocrop tidy, the understory bare.

Information and decision-making power now flowed straight to the top. Decades later when the first crop was felled, vast fortunes were made, tree by standardized tree. The clear-felled forests were replanted, with hopes of extending the boom. Readers of the American political anthropologist of anarchy and order, James C. Scott, know what happened next.

It was a disaster so bad that a new word, *Waldsterben*, or "forest death," was minted to describe the result. All the same species and age, the trees were flattened in storms, ravaged by insects and disease—even the survivors were spindly and weak. Forests were now so tidy and bare, they were all but dead. The first magnificent bounty had not been the beginning of endless riches, but a one-off harvesting of millennia of soil wealth built up by biodiversity and symbiosis. Complexity was the goose that laid golden eggs, and she had been slaughtered.

The story of German scientific forestry transmits a timeless truth: when we simplify complex systems, we destroy them, and the devastating consequences sometimes aren't obvious until it's too late.

That impulse to scour away the messiness that makes life resilient is what many conservation biologists call the "pathology of command and control." Today, the same drive to centralize, control, and extract has driven the internet to the same fate as the ravaged forests.

The internet's 2010s, its boom years, may have been the first glorious harvest that exhausted a onetime bonanza of diversity. The complex web of human interactions that thrived on the internet's initial technological diversity is now corralled into globe-spanning data-extraction engines making huge fortunes for a tiny few.

Our online spaces are not ecosystems, though tech firms love that word. They're plantations; highly concentrated and controlled environments, closer kin to the industrial farming of the cattle feedlot or battery chicken farms that madden the creatures trapped within.

We all know this. We see it each time we reach for our phones. But what most people have missed is how this concentration reaches deep into the internet's infrastructure—the pipes and protocols, cables and networks, search engines and browsers. These structures determine how we build and use the internet, now and in the future.

They've concentrated into a series of near-planetary duopolies. For example, as of April 2024, Google and Apple's internet browsers have captured almost 85 percent of the world market share, Microsoft and Apple's two desktop operating systems over 80 percent. Google runs 84 percent of global search and Microsoft 3 percent. Slightly more than half of all phones come from Apple and Samsung, while over 99 percent of mobile operating systems run on Google or Apple software. Two cloud computing providers, Amazon Web Services and Microsoft's Azure make up over 50 percent of the global market. Apple and Google's email clients manage nearly 90 percent of global email. Google and Cloudflare serve around 50 percent of global domain name system requests.

Two kinds of everything may be enough to fill a fictional ark and repopulate a ruined world, but can't run an open, global "network of networks" where everyone has the same chance to innovate and compete. No wonder internet engineer Leslie Daigle termed the concentration and consolidation of the internet's technical architecture "'climate change' of the Internet ecosystem."

Walled Gardens Have Deep Roots

The internet made the tech giants possible. Their services have scaled globally, via its open, interoperable core. But for the past decade, they've also worked to enclose the varied, competing, and often open-source or collectively provided services the internet is built on into their proprietary domains. Although this improves their operational efficiency, it also ensures that the flourishing conditions of their own emergence aren't repeated by potential competitors. For tech giants, the long period of open internet evolution is over. Their internet is not an ecosystem. It's a zoo.

Google, Amazon, Microsoft, and Meta are consolidating their control deep into the underlying infrastructure through acquisitions, vertical integration, building proprietary networks, creating chokepoints, and concentrating functions from different technical layers into a single silo of top-down control. They can afford to, using the vast wealth reaped in their one-off harvest of collective, global wealth.

Taken together, the enclosure of infrastructure and imposition of technology monoculture forecloses our futures. Internet people like to talk about "the stack," or the layered architecture of protocols, software, and hardware, operated by different service providers, that collectively delivers the daily miracle of connection. It's a complicated, dynamic system with a basic value baked into the core design: key functions are kept separate to ensure resilience and generality and to create room for innovation.

Initially funded by the U.S. military and designed by academic researchers to function in wartime, the internet evolved to work anywhere, in any condition, operated by anyone who wanted to connect. But what was a dynamic, ever-evolving game of Tetris with distinct "players" and "layers" is today hardening into a continent-spanning system of compacted tectonic plates. Infrastructure is not just what we see on the surface; it's the forces below, that make mountains and power tsunamis. Whoever controls infrastructure determines the future. If you doubt that, consider that in Europe we're still using roads and living in towns and cities the Roman Empire mapped out two thousand years ago.

In 2019, some internet engineers in the global standards-setting body, the Internet Engineering Task Force, raised the alarm. Daigle, a respected engineer who had previously chaired its oversight committee and internet architecture board, wrote in a policy brief that consolidation meant network structures were ossifying throughout the stack, making incumbents harder to dislodge and violating a core principle of the internet: that it does not create "permanent favorites." Consolidation doesn't just squeeze out competition. It narrows the kinds of relationships possible between operators of different services.

As Daigle put it: "The more proprietary solutions are built and deployed instead of collaborative open standards-based ones, the less the internet survives as a platform for future innovation." Consolidation kills collaboration between service providers through the stack by rearranging an array of different relationships—competitive, collaborative—into a single predatory one.

Since then, standards development organizations started several initiatives to name and tackle infrastructure consolidation, but these floundered. Bogged down in technical minutiae, unable to separate themselves from their employers' interests and deeply held professional values of simplification and control, most internet engineers simply couldn't see the forest for the trees.

Up close, internet concentration seems too intricate to untangle; from far away, it seems too difficult to deal with. But what if we thought of the internet not as a doomsday "hyperobject," but as a damaged and struggling ecosystem facing destruction? What if we looked at it not with helpless horror at the eldritch encroachment of its current controllers, but with compassion, constructiveness, and hope?

Technologists are great at incremental fixes, but to regenerate entire habitats, we need to learn from ecologists who take a whole-systems view. Ecologists also know how to keep going when others first ignore you and then say it's too late, how to mobilize and work collectively, and how to build pockets of diversity and resilience that will outlast them, creating possibilities for an abundant future they can imagine but never control. We don't need to repair the internet's infrastructure. We need to rewild it.

What Is Rewilding?

Rewilding "aims to restore healthy ecosystems by creating wild, biodiverse spaces," according to the International Union for Conservation of Nature. More ambitious and risk-tolerant than traditional conservation, it targets entire ecosystems to make space for complex food webs and the emergence of unexpected interspecies relations. It's less interested in saving specific endangered species. Individual species are just ecosystem components, and focusing on components loses sight of the whole. Ecosystems flourish through multiple points of contact between their many elements, just like computer networks. And like in computer networks, ecosystem interactions are multifaceted and generative.

Rewilding has much to offer people who care about the internet.

As Paul Jepson and Cain Blythe wrote in their book *Rewilding: The Radical New Science of Ecological Recovery*, rewilding pays attention "to the emergent properties of interactions between 'things' in ecosystems . . . a move from linear to systems thinking."

It's a fundamentally cheerful and workmanlike approach to what can seem insoluble. It doesn't micromanage. It creates room for "ecological processes [that] foster complex and self-organizing ecosystems." Rewilding puts into practice what every good manager knows: Hire the best people you can, provide what they need to thrive, then get out of the way. It's the opposite of command and control.

Rewilding the internet is more than a metaphor. It's a framework and plan. It gives us fresh eyes for the wicked problem of extraction and control, and new means and allies to fix it. It recognizes that ending internet monopolies isn't just an intellectual problem. It's an emotional one. It answers questions like: How do we keep going when the monopolies have more money and power? How do we act collectively when they suborn our community spaces, funding, and networks? And how do we communicate to our allies what fixing it will look and feel like?

Rewilding is a positive vision for the networks we want to live inside, and a shared story for how we get there. It grafts a new tree onto technology's tired old stock.

What Ecology Knows

Ecology knows plenty about complex systems that technologists can benefit from. First, it knows that shifting baselines are real.

If you were born around the 1970s, you probably remember many more dead insects on the windscreen of your parents' car than on your own. Global land-dwelling insect populations are dropping about 9 percent a decade. If you're a geek, you probably programmed your own computer to make basic games. You certainly remember a web with more to read than the same five websites. You may have even written your own blog.

But many people born after 2000 probably think a world with few insects, little ambient noise from birdcalls, where you regularly use only a few social media and messaging apps (rather than a whole *web*) is normal. As Jepson and Blythe wrote, shifting baselines are "where each generation assumes the nature they experienced in their youth to be normal and unwittingly accepts the declines and damage of the generations before." Damage is already baked in. It even seems natural.

Ecology knows that shifting baselines dampen collective urgency and deepen generational divides. People who care about internet monoculture and control are often told they're nostalgists harkening back to a pioneer era. It's fiendishly hard to regenerate an open and competitive infrastructure for younger generations who've been raised to assume that two or three platforms, two app stores, two operating systems, two browsers, one cloud/megastore, and a single search engine for the world comprise *the internet*. If the internet for you is the massive sky-scraping silo you happen to live inside and the only thing you can see outside is the single, other massive sky-scraping silo, then how can you imagine anything else?

Concentrated digital power produces the same symptoms that command and control produces in biological ecosystems: acute distress punctuated by sudden collapses once tipping points are reached. What scale is needed for rewilding to succeed? It's one thing to reintroduce wolves to the 3,472 square miles of Yellowstone, and quite another to cordon off about 20 square miles of a polder (land reclaimed from a body of water) known as Oostvaardersplassen near Amsterdam. Large and diverse Yellowstone is likely complex enough to adapt to change, but Oostvaardersplassen has struggled.

In the 1980s, the Dutch government attempted to regenerate a section of the overgrown Oostvaardersplassen. An independent-minded government ecologist, Frans Vera, said reeds and scrub would dominate unless now-extinct herbivores grazed them. In place of ancient aurochs, the state forest management agency introduced the famously bad-tempered German Heck cattle and in place of an extinct steppe pony, a Polish semi-feral breed.

Some thirty years on, with no natural predators, and after

plans for a wildlife corridor to another reserve came to nothing, there were many more animals than the limited winter vegetation could sustain. People were horrified by starving cows and ponies, and beginning in 2018, government agencies instituted animal welfare checks and culling.

Just turning the clock back was insufficient. The segment of Oostvaardersplassen was too small and too disconnected to be rewilded. Because the animals had nowhere else to go, overgrazing and collapse was inevitable, an embarrassing but necessary lesson. Rewilding is a work in progress. It's not about trying to revert ecosystems to a mythical Eden. Instead, rewilders seek to rebuild resilience by restoring autonomous natural processes and letting them operate at scale to generate complexity. But rewilding, itself a human intervention, can take several turns to get right.

Whatever we do, the internet isn't returning to old-school then-common interfaces like FTP and Gopher, or organizations operating their own mail servers again instead of off-the-shelf solutions like G Suite. But some of what we need is already here, especially on the web. Look at the resurgence of RSS feeds, email newsletters, and blogs, as we discover (yet again) that relying on one app to host global conversations creates a single point of failure and control. New systems are growing, like the Fediverse with its federated islands, or Bluesky with algorithmic choice and composable moderation.

We don't know what the future holds. Our job is to keep open as much opportunity as we can, trusting that those who come later will use it. Instead of setting purity tests for which kind of internet is most like the original, we can test changes against the values of the original design. Do new standards protect the network's "generality," i.e., its ability to support multiple uses, or is functionality limited to optimize efficiency for the biggest tech firms?

As early as 1985, plant ecologists Steward T.A. Pickett and Peter S. White wrote in *The Ecology of Natural Disturbance and Patch Dynamics* that an "essential paradox of wilderness conservation is that we seek to preserve what must change." Some internet engineers know this. David Clark, a Massachusetts Institute of Technology professor who worked on some of the internet's earliest protocols, wrote an entire book about other network architectures that might have been built if different values, like security or centralized management, had been prioritized by the internet's creators.

But our internet took off because it was designed as a general-purpose network, built to connect anyone.

Our internet was built to be complex and unbiddable, to do things we cannot yet imagine. When we interviewed Clark, he told us that "'complex' implies a system in which you have emergent behavior, a system in which you can't model the outcomes. Your intuitions may be wrong. But a system that's too simple means lost opportunities." Everything we collectively make that's worthwhile is complex and thereby a little messier. The cracks are where new people and ideas get in.

Internet infrastructure is a degraded ecosystem, but it's also a built environment, like a city. Its unpredictability makes it generative, worthwhile, and deeply human. In 1961, Jane Jacobs, an American Canadian activist and author of *The Death and Life of Great American Cities*, argued that mixed-use neighborhoods were safer, happier, more prosperous, and more livable than the sterile, highly controlling designs of urban planners like New York's Robert Moses.

Just like the crime-ridden, Corbusier-like towers Moses crammed people into when he demolished mixed-use neighborhoods and built highways through them, today's top-down, concentrated internet is, for many, an unpleasant and harmful place. Its owners are hard to remove, and their interests do not align with ours.

As Jacobs wrote: "As in all Utopias, the right to have plans of any significance belonged only to the planners in charge." As a top-down, built environment, the internet has become something that is done to us, not something we collectively remake every day.

Ecosystems endure because species serve as checks and balances on each other. They have different modes of interaction, not just extraction, but mutualism, commensalism, competition, and predation. In flourishing ecosystems, predators are subject to limits. They're just one part of a complex web that passes calories around, not a one-way ticket to the end of evolution.

Ecologists know that diversity is resilience.

On July 18, 2001, eleven carriages of a sixty-car freight train derailed in the Howard Street Tunnel under Mid-Town Belvedere, a neighborhood just north of downtown Baltimore. Within minutes, one carriage containing a highly flammable chemical was punctured. The escaping chemical ignited, and soon, adjacent carriages were alight in a fire that took about five days to put out. The disaster multiplied and spread. Thick, brick tunnel walls acted like an oven, and temperatures rose to nearly 2,000 degrees Fahrenheit. A more than three-foot-wide water main above the tunnels burst, flooding the tunnel with millions of gallons within hours. It only cooled a little. Three weeks later, an explosion linked to the combustible chemical blew out manhole covers located as far as two miles away.

WorldCom, then the second-largest long-distance phone company in the U.S., had fiber-optic cables in the tunnel carrying high volumes of phone and internet traffic. However, according to Clark, the MIT professor, WorldCom's resilience planning meant traffic was spread over different fiber networks in anticipation of just this kind of event.

On paper, WorldCom had network redundancy. But almost immediately, U.S. internet traffic slowed, and WorldCom's East Coast and transatlantic phone lines went down. The region's narrow physical topography had concentrated all those different fiber networks into a single chokepoint, the Howard Street Tunnel. WorldCom's resilience was, quite literally, incinerated. It had technological redundancy, but not diversity. Sometimes we don't notice concentration until it's too late.

Clark tells the story of the Howard Street Tunnel fire to show that bottlenecks aren't always obvious, especially at the operational level, and huge systems that seem secure, due to their size and resources, can unexpectedly crumble.

In today's internet, much traffic passes through tech firms' private networks, for example, Google's and Meta's own undersea cables. Much internet traffic is served from a few dominant content distribution networks, like Cloudflare and Akamai, who run their own networks of proxy servers and data centers. Similarly, that traffic goes through an increasingly small number of domain name system (DNS) resolvers, which work like phone books for the internet, linking website names to their numeric address.

All of this improves network speed and efficiency but creates new and non-obvious bottlenecks like the Howard Street Tunnel. Centralized service providers say they're better resourced and skilled at attacks and failures, but they are also large, attractive targets for attackers and possible single points of system failure.

On October 21, 2016, dozens of major U.S. websites suddenly stopped working. Domain names belonging to Airbnb, Amazon, PayPal, CNN, and *The New York Times* simply didn't resolve. All were clients of the commercial DNS service provider Dyn, which had been hit by a cyberattack. Hackers infected tens of thousands of internet-enabled devices with malicious software, creating a network of hijacked devices, or a botnet, that they used to bombard Dyn with queries until it collapsed. America's biggest internet brands were brought down by nothing more than a network of baby monitors, security webcams, and other consumer devices. Although they all likely had resilience planning and redundancies, they went down because a single chokepoint—in one crucial layer of infrastructure—failed.

Widespread outages due to centralized chokepoints have become so common that investors even use them to identify opportunities. When a failure by cloud provider Fastly took high-profile websites offline in 2021, its share price surged. Investors were delighted by headlines that informed them of an obscure technical service provider with an apparent lock on an essential service. To investors, this critical infrastructure failure doesn't look like fragility but like a chance to profit.

The result of infrastructural narrowness is baked-in fragility that we only notice after a breakdown. But monoculture is also highly visible in our search and browser tools. Search, browsing, and social media are how we find and share knowledge and how we communicate. They're a critical, global epistemic and democratic infrastructure, controlled by just a few U.S. companies. Crashes, fires, and floods may simply be entropy in action, but systemically concentrated and risky infrastructures are choices made manifest—and we can make better ones.

The Look & Feel of a Rewilded Internet

A rewilded internet will have many more service choices. Some services like search and social media will be broken up, as AT&T eventually was. Instead of tech firms extracting and selling people's personal data, different payment models will fund the infrastructure we need. Right now, there is little explicit provision for public goods like internet protocols and browsers, essential to making the internet work. The biggest tech firms subsidize and profoundly influence them.

Part of rewilding means taking what's been pulled into the big tech stack back out of it, and paying for the true costs of connectivity. Some things like basic connectivity we will continue to pay for directly, and others, like browsers, we will support indirectly but transparently, as described below. The rewilded internet will have an abundance of ways to connect and relate to each other. There won't be just one or two numbers to call if leaders of a political coup decide to shut the internet down in the middle of the night, as has happened in places like Egypt and Myanmar. No one entity will permanently be on top. A rewilded internet will be a more interesting, usable, stable, and enjoyable place to be.

Through extensive research, Nobel-winning economist Elinor Ostrom found that "when individuals are well informed about the problem they face and about who else is involved, and can build settings where trust and reciprocity can emerge, grow, and be sustained over time, costly and positive actions are frequently taken without waiting for an external authority to impose rules, monitor compliance, and assess penalties." Ostrom found people spontaneously organizing to manage natural resources—from water company cooperation in California to Maine lobster fishermen organizing to prevent overfishing.

Self-organization also exists as part of a key internet function: traffic coordination. Internet exchange points (IXPs) are an example of common-pool resource management, where internet service providers (ISPs) collectively agree to carry each other's data for low or no cost. Network operators of all kinds—telecoms companies, large tech firms, universities, governments, and broadcasters—all need to send large amounts of data through other ISPs' networks so that it gets to its destination.

If they managed this separately through individual contracts, they'd spend much more time and money. Instead, they often form IXPs, typically as independent, not-for-profit associations. As well as managing traffic, IXPs have, in many—and especially developing—countries, formed the backbone of a flourishing technical community that further drives economic development.

Both between people and on the internet, connections are generative. From technical standards to common-pool resource management and even to more localized broadband networks known as "altnets," internet rewilding already has a deep toolbox of collective action ready to be deployed.

The New Drive for Antitrust & Competition

The list of infrastructures to be diversified is long. As well as pipes and protocols, there are operating systems, browsers, search engines, the Domain Name System, social media, advertising, cloud providers, app stores, AI companies, and more. And these technologies also intertwined.

But showing what can be done in one area creates opportunities in others. First, let's start with regulation.

You don't always need a big new idea like rewilding to frame and motivate major structural change. Sometimes reviving an old idea will do. President Biden's 2021 "Executive Order on Promoting Competition in the American Economy" revived the original, pro-worker, trust-busting scope and urgency of the early twentieth-century legal activist and Supreme Court Justice Louis D. Brandeis, along with rules and framings that date back to before the 1930s New Deal.

U.S. antitrust law was created to break the power of oligarchs in oil, steel, and railroads who threatened America's young democracy. It gave workers basic protections and saw equal economic opportunity as essential to freedom. This view of competition as essential was whittled away by Chicago School economic policies in the 1970s and Reagan-era judges' court rulings over the decades. They believed intervention should only be permitted when monopoly power causes consumer prices to rise. The intellectual monoculture of that consumer-harm threshold has since spread globally.

It's why governments just stood aside as twenty-first-century tech firms romped to oligopoly. If a regulator's sole criterion for action is to make sure consumers don't pay a penny more, then the free or data-subsidized services of tech platforms don't even register. (Of course, consumers pay in other ways, as these tech giants exploit their personal information for profit.) This laissez-faire approach allowed the biggest firms to choke off competition by acquiring their competitors and vertically integrating service providers, creating the problems we have today.

Regulators and enforcers in Washington and Brussels now say they have learned that lesson and won't allow AI dominance to happen as internet concentration did. Federal Trade Commission Chair Lina Khan and U.S. Department of Justice antitrust enforcer Jonathan Kanter are identifying chokepoints in the AI "stack"—concentration in control of processing chips, datasets, computing capacity, algorithm innovation, distribution platforms, and user interfaces—and analyzing them to see if they affect systemic competition. This is potentially good news for people who want to prevent the current dominance of tech giants being grandfathered into our AI future.

In his 2021 signing of the executive order on competition, President Biden said: "Capitalism without competition isn't capitalism; it's exploitation." Biden's enforcers are changing the kinds of cases they take up and widening the applicable legal theories on harm that they bring to judges. Instead of the traditionally narrow focus on consumer prices, today's cases argue that the economic harms perpetrated by dominant firms include those suffered by their workers, small companies, and the market as a whole.

Khan and Kanter have jettisoned narrow and abstruse models

of market behavior for real-world experiences of health-care workers, farmers, and writers. They *get* that shutting off economic opportunity fuels far-right extremism. They've made antitrust enforcement and competition policy explicitly about coercion versus choice, power versus democracy. Kanter told a recent conference in Brussels that "excessive concentration of power is a threat . . . it's not just about prices or output but it's about freedom, liberty and opportunity."

Enforcers in Washington and Brussels are starting to preemptively block tech firms from using dominance in one realm to take over another. After scrutiny by the U.S. FTC and European Commission, Amazon recently abandoned its plan to acquire the home appliance manufacturer iRobot. Regulators on both sides of the Atlantic have also moved to stop Apple from using its iPhone platform dominance to squeeze app store competition and dominate future markets through, for example, pushing the usage of CarPlay on automakers and limiting access to its tap-to-pay digital wallet in the financial services sector.

Still, so far, their enforcement actions have focused on the consumer-facing, highly visible parts of the tech giants' exploitative and proprietary internet. The few, narrow measures of the 2021 executive order that aim to reduce infrastructure-based monopolies only prevent *future* abuses like radio spectrum-hogging, not those already locked in. Sure, the best way to deal with monopolies is to stop them from happening in the first place. But unless regulators and enforcers eradicate the existing dominance of these giants now, we'll be living in today's infrastructure monopoly for decades, perhaps even a century.

Even activist regulators have shied away from applying the toughest remedies for concentration in long-consolidated markets, such as non-discrimination requirements, functional interoperability, and structural separations, i.e., breaking companies up. And declaring that search and social media monopolies are actually public utilities—and forcing them to act as common carriers open to all—is still too extreme for most.

But rewilding a built environment isn't just sitting back and seeing what tender, living thing can force its way through the concrete. It's razing to the ground the structures that block out light for everyone not rich enough to live on the top floor.

When the writer and activist Cory Doctorow wrote about how to free ourselves from the clutches of Big Tech, he said that though breaking up big companies will likely take decades, providing strong and mandatory interoperability would open up innovative space and slow the flow of money to the largest firms—money they would otherwise use to deepen their moats.

Doctorow describes "comcom," or competitive compatibility, as a kind of "guerrilla interoperability, achieved through reverse engineering, bots, scraping and other permissionless tactics." Before a thicket of invasive laws sprung up to strangle it, comcom was how people figured out how to fix cars and tractors or rewrite software. Comcom drives the try-every-tactic-until-one-works behavior you see in a flourishing ecosystem.

In an ecosystem, diversity of species is another way of saying "diversity of tactics," as each successful new tactic creates a new niche to occupy. Whether it's an octopus camouflaging itself as a sea snake, a cuckoo smuggling her chicks into another bird's nest, orchids producing flowers that look just like a female bee, or parasites influencing rodent hosts to take life-ending risks, each evolutionary micro-niche is created by a successful tactic. Comcom is simply tactical diversity; it's how organisms interact in complex, dynamic systems. And humans have demonstrated the epitome of short-term thinking by enabling the oligarchs who are trying to end it.

Efforts are underway. The EU already has several years of experience with interoperability mandates and precious insight into how determined firms work to circumvent such laws. The U.S., however, is still in its early days of ensuring software interoperability, for example, for videoconferencing.

Perhaps one way to motivate and encourage regulators and enforcers everywhere is to explain that the subterranean architecture of the internet has become a shadowland where evolution has all but stopped. Regulators' efforts to make the visible internet competitive will achieve little unless they also tackle the devastation that lies beneath.

Next Steps

Much of what we need is already here. Beyond regulators digging deep for courage, vision, and bold new litigation strategies, we need vigorous, pro-competitive government policies around procurement, investments, and physical infrastructure. Universities must reject research funding from tech firms because it always comes with conditions, both spoken and unspoken.

Instead, we need more publicly funded tech research with publicly released findings. Such research should investigate power concentration in the internet ecosystem and practical alternatives to it. We need to recognize that much of the internet's infrastructure is a de facto utility that we must regain control of.

We must ensure regulatory and financial incentives and support for alternatives including common-pool resource management, community networks, and the myriad other collaborative mechanisms people have used to provide essential public goods like roads, defense, and clean water.

All this takes money. Governments are starved of tax revenue by the once-in-history windfalls seized by today's tech giants, so it's clear where the money is. We need to get it back.

We know all this, but still find it so hard to collectively act. Why?

Herded into rigid tech plantations rather than functioning, diverse ecosystems, it's tough to imagine alternatives. Even those who can see clearly may feel helpless and alone. Rewilding unites everything we know we need to do and brings with it a whole new toolbox and vision.

Ecologists face the same systems of exploitation and are organizing urgently, at scale and across domains. They see clearly that the issues aren't isolated but are instances of the same pathology of command and control, extraction, and domination that political anthropologist James Scott first noticed in scientific forestry. The solutions are the same in ecology and technology: aggressively use the rule of law to level out unequal capital and power, then rush in to fill the gaps with better ways of doing things.

Keep the Internet, the Internet

Susan Leigh Star, a sociologist and theorist of infrastructure and networks, wrote in her 1999 influential paper, "The Ethnography of Infrastructure":

> Study a city and neglect its sewers and power supplies (as many have), and you miss essential aspects of distributional justice and planning power. Study an information system and neglect its standards, wires, and settings, and you miss equally essential aspects of aesthetics, justice, and change.

The technical protocols and standards that underlie the internet's infrastructure are ostensibly developed in open, collaborative standards development organizations (SDOs), but are also increasingly under the control of a few companies. What appear to be "voluntary" standards are often the business choices of the biggest firms.

The dominance of SDOs by big firms also shapes what does *not* get standardized—for example, search, which is effectively a global monopoly. While efforts to directly address internet consolidation have been raised repeatedly within SDOs, little progress has been made. This is damaging SDOs' credibility, especially outside the U.S. SDOs must radically change or they will lose their implicit global mandate to steward the future of the internet.

We need internet standards to be global, open, and generative. They're the wire models that give the internet its planetary form, the gossamer-thin but steely-strong threads holding together its interoperability against fragmentation and permanent dominance.

Make Laws & Standards Work Together

In 2018, a small group of Californians maneuvered the legislature into passing the California Consumer Privacy Act. Nested in the statute was an unassuming provision, the "right to opt out of sale or sharing" your personal information via a "user-enabled global privacy control" or GPC signal that would create an automated method for doing so. The law didn't define how GPC would work. Because a technical standard was required for browsers, businesses, and providers to speak the same language, the signal's details were delegated to a group of experts.

In July 2021, California's attorney general mandated that all businesses use the newly created GPC for California-based consumers visiting their websites. The group of experts is now shepherding the technical specification through global web standards development at the World Wide Web Consortium. For California residents, GPC automates the request to "accept" or "reject" sales of your data, such as cookie-based tracking, on its websites. However, it isn't yet supported by major default browsers like Chrome and Safari. Broad adoption will take time, but it's a small step in changing real-world outcomes by driving anti-monopoly practices deep into the standards stack—and it's already being adopted elsewhere.

GPC is not the first legally mandated open standard, but it was deliberately designed from day one to bridge policymaking and standards-setting. The idea is gaining ground. A recent United Nations Human Rights Council report recommends that states delegate "regulatory functions to standard-setting organizations."

Make Service Providers—Not Users—Transparent

Today's internet offers minimal transparency of key internet infrastructure providers. For example, browsers are highly complex pieces of infrastructure that determine how billions of people use the web, yet they are provided for free. That's because the most commonly used search engines enter into opaque financial deals with browsers, paying them to be set as the default. Since few people change their default search engine, browsers like Safari and Firefox make money by defaulting the search bar to Google, locking in its dominance even as the search engine's quality of output declines.

This creates a quandary. If antitrust enforcers were to impose competition, browsers would lose their main source of income. Infrastructure requires money, but the planetary nature of the internet challenges our public funding model, leaving the door open to private capture. However, if we see the current opaque system as what it is, a kind of non-state taxation, then we can craft an alternative.

Search engines are a logical place for governments to mandate the collection of a levy that supports browsers and other key internet infrastructure, which could be financed transparently under open, transnational, multi-stakeholder oversight.

Make Space to Grow

We need to stop thinking of internet infrastructure as too hard to fix. It's the underlying system we use for nearly everything we do. The former prime minister of Sweden Carl Bildt and former Canadian deputy foreign minister Gordon Smith wrote in 2016 that the internet was becoming "the infrastructure of all infrastructure." It's how we organize, connect, and build knowledge, even—perhaps—planetary intelligence. Right now, it's concentrated, fragile, and utterly toxic.

Ecologists have reoriented their field as a "crisis discipline," a field of study that's not just about learning things but about saving them. We technologists need to do the same. Rewilding the internet connects and grows what people are doing across regulation, standards-setting, and new ways of organizing and building infrastructure, to tell a shared story of where we want to go. It's a shared vision with many strategies. The instruments we need to shift away from extractive technological monocultures are at hand or ready to be built.

RIVKA GALCHEN

Pecking Order

FROM *The New Yorker*

ON A DRIZZLY day in Grünau im Almtal, Austria, a gaggle of greylag geese shared a peaceful moment on a grassy field near a stream. One goose, named Edes, was preening quietly; others were resting with their beaks pointed tailward, nestled into their feathers. Then a camouflaged speaker that scientists had placed nearby started to play. First came a recorded honk from an unpartnered male goose named Joshua. Edes went on with his preening. Next came a honk that was lower in pitch than the first, with a slight bray. Edes looked up. As the other geese remained tucked in their warm positions, incurious, Edes scanned the field. He had just heard a recorded "distance call" from his life partner, a female goose whom scientists had named Bon Jovi.

Edes and his fellow geese live near the Konrad Lorenz Research Center for Behavior and Cognition, which is named for a Nobel laureate whose imprinting experiments, in the 1930s, convinced goslings that he was their mother. (They took to following him in a downy line.) Greylag geese in the area have been studied continually ever since. The director of the center, a biologist and bird ecologist named Sonia Kleindorfer, showed me footage of Edes to demonstrate the subtlety of goose communication.

Geese maintain elaborate social structures, travel in family groups, and can navigate from Sweden to Spain. In a fight, an unpartnered greylag goose has a higher heart rate than a partnered one, and the heart rate of a recently widowed goose can remain depressed for about a year. These birds have things to discuss. Still, geese are not the Ciceros of the bird world. A lyrebird sings long, elaborate songs; ravens really can say "nevermore." Geese are known for nasal honks. How much nuance can there be in a honk?

Greylag geese, it turns out, have at least ten different kinds of calls. "We are completely underappreciating the way they communicate," Kleindorfer told me. "They give a departure call when they leave, and a contact call after they arrive. They know if their allies are there, if the bold geese are there. There is so much information that geese are getting from calls."

Bird vocalizations are usually divided into songs and calls, but these are wobbly categories. What is designated a song in one species may be shorter in duration than what, in another species, is termed a call. Onomatopoeic groupings such as *tseets*, *chirrups*, *rreeyoos*, *seeew-soooos*, and *dahs* are also indeterminate: people transcribe the same sounds in different ways, and no bird version of the Académie Française exists to adjudicate. The vocalizations of birds are fundamentally incommensurate with human ones. We have a larynx and two vocal cords; they have what's called a syrinx, which is a bit like having two larynxes that you can use at the same time.

Kleindorfer, the daughter of a mathematician and an actress, looks like a cross between a hiker and the film star Sophia Loren. From February to April, she researches Darwin's finches in the Galápagos; from September to December, songbirds in Australia; and, for the rest of the year, the geese outside her office door. Early in her education, as an undergraduate at the University of Pennsylvania, she was taught that "male songbirds sing, females don't, and if females do sing it's an error." The attitude at the time, she told me, was that "females are drab, inconspicuous, and quiet." A few years after earning a PhD in zoology at the University of Vienna, Kleindorfer took a job as a research biologist at Flinders University, in Australia, where songbird species originally evolved. "Imagine my surprise," she told me. "I heard all these females singing songs as complex as the male songs." Much of her ensuing career has focused on bird vocalizations that were either underappreciated or unknown.

Kleindorfer decided to study bird eggs and early development, which were then neglected research topics. "Maybe this was because only females have eggs and I was a woman in science," she told me. "I don't have a better reason." Kleindorfer had noticed that mustached-warbler chicks seemed to respond to the alarm calls of adult warblers, even though the thinking at the time was that such calls were directed at other adults, or possibly at predators. "If I put a snake nearby, the parental alarm call made the chicks in the nest jump," she said. "If I put a marsh harrier"—a hawklike predatory bird—"nearby, the response to the parental alarm call was that the chicks would duck." The chicks were responding appropriately to different alarm calls—a satisfying finding.

Kleindorfer also studied the superb fairy wren, a songbird that weighs about as much as a walnut and sports a flirty, upright tail. Despite their fanciful names, fairy wrens are commonplace in Australia. They are socially monogamous but sexually promiscuous—they are essentially in open marriages—and they bring up their young collectively. Arguably, they have even more to chat about than geese do. Fairy wren nests are about the size of cupped human hands, built to contain pale, speckled eggs that are smaller than thumbnails. Kleindorfer and her team wired up nests with cameras and microphones and soon discovered something that they hadn't known to look for. "The mothers in nests were producing an incubation call—a call to the eggs," she told me. It was like a lullaby. Why would a mother bird make any sound that could attract predators to the nest? "Songbird embryos don't have well-developed ears, so this was completely unexpected," she said. "That started a twenty-year project—why is she calling to the eggs?"

The team compared incubation calls to the begging calls of young chicks. "It was very odd," Kleindorfer recalled. "Each nest had its own distinct begging call." What's more, each begging call matched an element from the mother's incubation call.

This suggested, startlingly, that birds could learn a literal mother tongue while still in ovo. (Humans do this too; French and German babies have distinct cries.) Even "foster" chicks, who as eggs were physically moved from one nest to another, learned begging calls from their foster mothers, rather than from their genetic mothers. This was big news in the ornithology world. "The paradigm of how songbirds learn—after hatching, from their father's song—was overthrown," she said. The same process was soon documented in more songbird species.

Language is often cited as the quality that distinguishes us as humans. When I asked Robert Berwick, an MIT computational linguist, about birds, he argued that "they're not trying to *say* anything in the sense of James Joyce trying to say something." Still, he and Kleindorfer both pointed out that humans and songbirds share a trait that many animals lack: we are "vocal learners," meaning that we can learn to make new sounds throughout our lives. (Bats, whales, dolphins, and elephants can too.) "To me, the most amazing thing is that every generation of vocal learners has its own sound," Kleindorfer said. "So, just like our English is different from Shakespeare's English, the songbirds, too, have very different songs from five hundred years ago. I am sure of it." We humans have long tried, often mistakenly, to differentiate ourselves from nonhuman animals—by arguing that only we have souls, or use tools, or are capable of self-awareness. Perhaps we should see what the birds have to say.

Animals have prominent speaking roles in many of our oldest stories. Eve has a memorable conversation with a snake. In Norse mythology, two ravens, Huginn and Muninn, serve as spies to the god Odin, whispering to him the news of the world. In many cultures, the "language of birds" refers to a divine or perfect language—the language of angels. In the scientific realm, however, the notion that nonhuman animals use language is often seen as foolish or naive. Some birds may be excellent mimics, like parrots, but they can also mimic chainsaws or barking dogs; scholars don't usually consider imitation a form of understanding. The prevailing dogma is that birds sing either to impress mates or to defend their territory. (I suspect that most of human communication could also be slotted into those categories.) In college, I was taught a stranger but similarly diminishing idea: that songbirds sing in the morning to burn fat, so that they are light enough to fly around during the day. Apparently, this idea is no longer taken seriously.

Even among species we view as being closer to ourselves, such as primates, scientists have tended to talk about "communication" instead of "language." But it's tricky to say where the line is, or what we mean by "communication," since even bacteria communicate, as Berwick pointed out to me. "I think it's best to think of language not as speech but as a cognitive ability in the mind that sometimes leads to speech," he said, giving the example of inward conversations we have with ourselves. The linguist Noam Chomsky has said, "It's about as likely that an ape will prove to have a language ability as there is an island somewhere with a species of flightless birds waiting for humans to teach them to fly." Chomsky's 2017 book on the evolution of language, co-authored with Berwick, is titled *Why Only Us*.

Over the years, however, some researchers have looked closely at the contexts in which certain animal vocalizations are made. In the late 1970s, two primatologists, Dorothy Cheney and Robert Seyfarth, were studying vervet monkeys in Kenya. Vervets have dark faces and pale fur; they are about the size of a small backpack and are hunted by pythons, eagles, and leopards. Cheney and Seyfarth documented something remarkable: one recorded vervet vocalization made vervets look up, presumably for eagles; another made them look down, presumably for pythons; and a third sent them running up into the trees, a good defense against approaching leopards. Young vervets sometimes use these calls faultily, perhaps sounding a leopard alarm for a warthog. But they get better as they grow up. They learn.

A newer generation of scientists has been trying to understand bird vocalizations. The alarm calls of Siberian jays can be said to have been partially translated. One of their screeches indicates a sitting hawk (which prompts other jays to come together in a group), another a flying hawk (jays hide, which makes them difficult to spot), and a third a hawk actively attacking (jays fly to the treetops to search for the attacker, and possibly flee).

When cheery birds known as tufted titmice make a piercing sound, other titmice may respond by collectively harrying an invading predator. Some birds even lie. Fork-tailed drongos—common, innocuous-looking little dark birds that live in Africa—sometimes mimic the alarm calls of starlings or meerkats. Duped listeners flee the nonexistent threat, leaving behind a buffet for the drongo.

Upon seeing an owl, a chickadee might sound a loud *chick-a-dee-dee-dee*, adding *dees* in relation to how dangerous the predator is perceived to be. This call is also understood by nuthatches, which will join in to mob and harass the predator, forming a kind of defensive alliance. If you record an Australian bird warning of a nearby cuckoo—cuckoos leave their eggs in the nests of other species and often kill their step-siblings—birds in China will understand the call.

Kleindorfer considers Cheney and Seyfarth, the primatologists, to be important sources of inspiration. After she moved to Australia, she and her colleagues built up a sound library of Australian songbirds. They also made recordings of quieter, familiar bird sounds, such as the incubation calls. Each family unit, they discovered, had its own "familect," a system of sounds that chicks learn from their parents. Curiously, chicks seemed to adopt sounds sung either by their mother or by their father—but they avoided the sounds used by both parents. If the mother sings ABCXYZ and the father sings ABCGHI, then the chicks tend to sing the sound units X, Y, Z, G, H, and I. It's as if the young birds separate themselves from their parents by not speaking the shared sounds, but also stay close to their parents by learning what's unique to Mom and unique to Dad. When female chicks grow up, they are attracted to mates whose repertoire is familiar (he's one of us!), but not too familiar (he's not my brother or dad).

Birds in general are turning out to have intellectual abilities far greater than most people had imagined. It's not just that parrots and crows can do math as ably as young children, or that scrub jays cleverly cache and then uncache their food to fool other jays. Even inconspicuous and uncelebrated birds are capable of learning, and of sharing their learning with others. In the 1920s, tits from Swaythling, England, figured out how to open the caps of milk bottles, and by the late '40s tits across Ireland, Wales, and England had learned the trick. If language is more a capacity than it is a speech act, it seems possible that birds possess it.

In 1889, Ludwig Paul Koch, an eight-year-old boy in Frankfurt, Germany, received a present from his father: an Edison phonograph and some wax cylinders for recording sounds. The oldest known audio of birdsong is young Koch's recording of his pet white-rumped shama, a smallish songbird with a dark head, an orange body, and feathers that resemble a white bustle on its glossy black tail. A shama sings like a small chamber orchestra, with slippery, percussive, and sweet sounds in phrases of varying lengths. Many similar recordings followed. In 1929, the Cornell Library of Natural Sounds—now the Macaulay Library—was started with a few hard-won recordings of a sparrow, a wren, and a grosbeak. (Cornell is to ornithology what the Juilliard School is to music.)

Koch, who was Jewish, became a professional musician but fled Germany in the 1930s. In England, he became a beloved presence on BBC radio. Sounding like a singsong, sanguine sibling of Werner Herzog, he guided Brits through the charms of birdsong. ("A yellow icterine warbler," he told listeners, "frequently called me by my Christian name . . . Ludwig, Ludwig.") Koch often expressed the hope that such recordings might be used for science. Many years later, they were.

In 2010, Grant Van Horn was an undergraduate at UC San Diego and working in a computer-vision and machine-learning lab led by the computer scientist Serge Belongie. The lab was looking for a good data set to train an image-recognition program. At the time, Van Horn told me, many of the top images on Flickr, the popular photography website, were of birds. Van Horn was no birder, but he wanted to see if he and his colleagues could teach a computer program to distinguish between closely related species, such as a house wren and a marsh wren. As it turned out, they could.

The lab's work soon attracted the attention of ornithology researchers at Cornell. Van Horn recalls them telling him and his colleagues, "in the nicest possible way, 'Look, guys—this data set is quaint and poorly constructed, and the species that you chose to study make no sense. Do you want to effectively redo this whole process, but do it in collaboration with us?'" When Van Horn visited Cornell, the scientists took him out birding every morning and evening, and he remembers wishing that he could take their expertise back to California with him. The collaboration eventually helped the Cornell Lab of Ornithology develop Merlin Bird ID, an app that could reliably identify several hundred species of bird from photographs. It proved immensely popular—but the Cornell team had always had larger ambitions. "They kept asking, 'How can we do this with sound?'" Van Horn recalled. That was what the scientists were most interested in. But he assumed that auditory recognition was outside his expertise—until, out of curiosity, he attended a workshop on audio-related machine learning.

"I kind of had an epiphany," Van Horn said. Sound recognition often relied on spectrograms, visual representations of sounds similar to what you see in audio-editing software. Mike Webster, an animal-communication expert at Cornell who directs the Macaulay Library, and who worked on Merlin, told me, "When people figured out how to visualize sound—how to actually take measures of it—that led to just an explosion of research and understanding about how and why birds communicate with each other." Much of the work in sound recognition, Van Horn realized, was actually visual: "I thought, Let me bring these computer-vision skills to bear." An early test could differentiate between recordings of alder flycatchers and willow flycatchers.

In 2021, the number of bird recordings in the Macaulay Library, many of which were submitted by citizen scientists, reached a million. That same year, Cornell released Sound ID in the Merlin Bird ID app, which was originally trained on around 250 hours of bird sounds, as well as on background noises (whistling wind, passing cars), all manually annotated by experts. At first, Sound ID could identify about three hundred different North American birds, with a bias toward those found around Cornell. Three years and a million additional recordings later, Sound ID can now very accurately identify about fourteen hundred species.

The lab hopes that number can grow to roughly eight thousand, out of around eleven thousand known bird species.

Amateurs now have a remarkable ability to recognize the birds that are cooing or chirping—which has generated more interest in birds and directed more citizen-science recordings to the Macaulay Library. But decoding the bird vocalizations is another matter. One problem is that certain sorts of recordings are more plentiful than others. "Most of our database is songs," Webster said. "We can now understand songs at a level that we couldn't before." Alarm calls are also relatively easy to capture. But something like nest chatter, which is quieter and less predictable, is more elusive: "There are whole categories of bird communication that we've hardly even started to look at." Webster isn't expecting there to be straightforward translations of birds' sounds into human language; animals live in perceptual worlds that are just too different from our own. Still, he sees machine learning as a powerful new tool. "There are a lot of people who have dreams of using AI to allow us to decipher what animals are saying," he told me.

After three decades of research, Webster is preparing to retire. When I asked him what he hoped the next generation of scientists might learn, he thought for a moment. "Well, social birds. They are constantly chatting to each other," he said. "Making little noises. Often very quietly. It's like they're having a whisper conversation. What in the hell are they saying to one another? I'd really like to know."

Until my eleven-year-old daughter became interested in birds, I barely knew a starling from a sparrow. She once asked me, incredulously, "You're saying you can't tell a male sparrow from a female?" For a long time, we lived just east of the Lincoln Tunnel, where "birds" meant pigeons and seagulls, but within weeks of moving to Brooklyn we saw a red-tailed hawk on a lamppost. My daughter began talking about dark-eyed juncos and tufted titmice and peregrine falcons; we started visiting bird sanctuaries, and I eventually outgrew my favoritism toward mammals. Like millions of others, we started to use Merlin Bird ID. Usually, we heard birds before we saw them. Some local sparrows nesting in a hollow pole on our block sounded like Laurel and Hardy bickering.

"Anthropomorphism" is a familiar term that describes a common error: the assumption that animals have human qualities. A less familiar term, "anthropectomy," also describes a kind of error—that of baselessly assuming animals *don't* share certain qualities with us. Which kind of error is a person more likely to make? Or are these not errors but, rather, starting points, with someone like Jane Goodall starting from the premise that 98.7 percent of our DNA is shared with chimps, and someone else starting from the fact that we humans have sequenced our own DNA and no other species has even invented pliers? Since we're still arguing about what language is, it's difficult to say which assumption about animal language is more presumptuous.

Toshitaka Suzuki first started to wonder if birds speak their own language during his last year of college, at Toho University, in Tokyo. He was on a hike in the forests of Karuizawa when he witnessed what struck him as a strange drama among some common Japanese tits, birds that resemble chickadees. One tit called out *dee-dee-dee* near some scattered sunflower seeds; other tits flew over and began to eat. "Then one bird called out *hee-hee-hee*, and the birds all flew off into nearby bushes," Suzuki recalled. He could see no reason for them to abandon their feast.

Seconds later, a sparrow hawk swooped in; all of the tits had escaped safely. "I thought, Maybe *hee-hee-hee* means 'Hawk incoming, run away!'" Suzuki said. He already took birds seriously and knew a lot about them; he had studied under Hiroshi Hasegawa, a scientist who was central in bringing the short-tailed albatross back from near extinction. But Suzuki had thought of bird vocalizations as, for the most part, emotive, like music, or as a kind of beautiful nonsense. He has now devoted eighteen years to researching tits and their communication. "I couldn't have imagined how long I would be studying tits, because I love other animals as well," Suzuki told me. Like many researchers, he hopes that the more we understand birds the likelier we are to protect them.

In April 2023, at the University of Tokyo, Suzuki founded what he calls the world's first laboratory specifically devoted to animal linguistics. He argues that more work should be done to explore what cognitive abilities underlie human language—and then to investigate whether these abilities are present in animals. (In many ways, this approach mirrors the work of Berwick and Chomsky, but leads to different conclusions.) Some are skeptical of his push to compare animal communication to human language. "It's just so far removed from the complexity of human language that it doesn't make sense to use the same word," Todd Freeberg, an animal-communication researcher at the University of Tennessee, told me. When a chickadee amplifies a call by adding *dees*, some researchers might say that they are engaging in referential signaling, by adjusting their call to the seriousness of the threat. But Freeberg points out that extra *dees* could also be a result of heightened arousal, in general—less a conscious message than a physical response.

Like Kleindorfer, Suzuki took an interest in nests. Early on, he showed that chicks in nesting boxes respond to a call associated with crow sightings by crouching, and to a call associated with snakes by fleeing the nest altogether. The arc of Suzuki's research has, to some extent, followed a series of arguments about what qualities are required for communication to rise to the status of language. Humans are noted for their ability to form a mental image—a concept—of what they are communicating. Suzuki designed an experiment in which he played a variety of calls and moved a stick in a variety of ways; only when he played a snake-alarm call and moved a stick in a snakelike way did the birds tend to react as if a snake were present. To him, this suggested that they had some concept of snake-ness. (He said that the experiment was inspired by the way that humans perceive shapes in clouds.)

In a 2023 study, Suzuki showed that tits responded differently to a recorded ABCD call than they did to a remixed version of the call, such as DABC—a potential challenge to linguists who see sophisticated syntax as being unique to human language. (Studies of southern pied babblers and of chestnut-crowned babblers also have interesting syntax results.) Symbolic gestures—also often considered unique to humans—were addressed in a particularly adorable Suzuki paper, in 2024. His team watched mated pairs of tits as they entered their nest boxes. The opening to each nest box was small, allowing only one bird at a time to pass. But sometimes one bird, usually the female, fluttered its wings in what seemed to be an "after you" gesture. The other bird would then enter the box first. The fluttering didn't point at the nest box. In Suzuki's view, this suggested that the flutter was not a simple indication but, instead, a symbolic gesture—another item crossed off on the unique-to-human-language list.

Perhaps the nest-box study needs to be replicated; perhaps there are alternative interpretations of the results in the concept and syntax studies. Suzuki is open to such critiques. But he is also skeptical of many prominent ideas in linguistics, such as Chomsky and Berwick's argument that a slight evolutionary change in the brain unlocked a new linguistic capacity in humans: the unique and powerful ability to connect individual units in a hierarchical and expressive way. (Suzuki thinks that language more likely emerged bit by bit.) By Suzuki's latest count, the tit's vocal repertoire has more than two hundred distinct calls and phrases. He has many more experiments to conduct.

Recently, my daughter and I took an early-morning trip to Little Stony Point, in the Hudson Valley, and met up with two people who have no particular need for an app like Merlin Bird ID. Andrew C. Vallely does field-ornithology work for the American Museum of Natural History; he's become friends, by way of bird-watching, with Jeffrey Yang, an editor at New Directions Publishing. Yang had been seeing a lot of migrating warblers, which had flown well over a thousand miles—did we want to come try our luck? At his suggestion, I warned my daughter that there was no telling whether we'd actually see any.

About five minutes down the trail, in a not particularly distinguished wood (we could still hear cars and an excavator across the river), we saw a kingfisher diving and an adolescent eagle on a bare tree. As we walked, stopped, walked, stopped, we repeatedly heard what sounded like the call of a red-tailed hawk. But it soon became clear, at least to Vallely and Yang, that it was a jay mimicking a hawk. "They do that sometimes," Vallely told me.

"To scare away other predators?"

"That's one thought," he said. "Vocal mimicry can be pretty mysterious."

We heard the *tea kettle tea kettle* call of a Carolina wren; it sounded like a game of marbles to me. We saw a warbling vireo, a Cape May warbler, a blackpoll warbler, and a black-and-white warbler—birds so small that it was difficult to fathom how far some of them had traveled to be there. We heard little chips that sounded like a window being cleaned; a crickety decrescendo that was not made by crickets; a sound like a trill running into a wall; a high-pitched three-fast-one-slow, like a child playing Beethoven's Fifth Symphony. We encountered forty-four species by Yang's able count, and at the very end we saw a Swainson's thrush, who apparently wasn't in the mood to show off. Bird-watching, I thought, is a misleading term. So much of the fleeting, present-tense pleasure of it is bird-listening.

The quiet of the pandemic brought natural sounds to the foreground for Maddie Cusimano, who was then a graduate researcher of auditory perception at MIT's Center for Brains, Minds, and Machines. "Like a lot of people, I had the sense of getting to know the birds around me for the first time," she told me. Two doves were often visible from her window; she read that, in some dove pairs, one bird sings to the other in the mornings, and in the evenings the roles reverse. Cusimano was familiar enough with machine learning that, when she tried out bird-identification apps, she thought, I could help make this kind of thing.

Cusimano is now a senior scientist specializing in AI research at the Earth Species Project (ESP), a nonprofit dedicated to "using artificial intelligence to decode non-human communication." ESP's current efforts examine such species as zebra finches, crows, and beluga whales, but its early work has been preoccupied with preliminary challenges: the "cocktail-party problem" of picking up individual sounds in a noisy environment; how to correlate particular noises with the precise contexts in which they occur. "It's like we want to write the Magna Carta, but first we need to make the quill," Katie Zacarian, the organization's co-founder and CEO, told me. Zacarian isn't expecting a Google Translate for animal languages, but she does believe that we can understand animals better. She remembers that, as a kid, people often brought her father, an entomologist, pictures and specimens and asked: What is this? Her mother was a researcher and an administrator of multilingual education programs. "There's this underlying current, in their work, of decoding," she told me.

When I talked to Cusimano, on Zoom, she pulled up a collection of sound files of crows. Her data set comes from Daniela Canestrari and Vittorio Baglione, researchers at the University of León, in Spain, who have been studying Spanish carrion crows for more than twenty-five years. Cusimano has spent countless hours sitting in San Francisco, listening to these birds, and can sometimes guess which one she's hearing. "This is maybe what you expect a crow to sound like," she said, playing me two caws. "But then there's also this," she said, playing a whispery rasp. "Some sounds are very long." She played a ghostly *oooo*. "Then these two sounds, which you would never think were coming from crows." One sounded like the click of a computer mouse; another sounded froggy. Her favorite recording reminded me of a duck's quack. "I love these sounds," she told me.

One ambition of Cusimano's work is to find correlations between these varied vocalizations and the precise contexts in which they occurred. Research partners recently identified a quiet grunt that is most often made right when an adult crow returns to a nest—perhaps a way of saying, "Wake up!" A small biologger on the back of a crow can provide audio along with other data—a bird's-tail view. "You hear their wingbeats, you can hear their friends calling, and them calling back to their friends," Cusimano told me. The data feels intimate: "You hear baby chicks a distance away, then you hear the bird take off, and the chick sounds are getting louder. And then the crow lands in the nest. You're in the middle of this crow family."

Can a machine be trained to distinguish individual birds by the sounds they make? Can it pick up on vocalizations across individuals which share similar functions? Machine learning is excellent at detecting correlations, but some are irrelevant and even misleading. Cusimano developed an algorithm to distinguish among caws made by various crows, which had names such as Naranja, Rosa, and Azul. She seemed to have succeeded. Then she realized that the computer might be categorizing the sounds based on distinctive background noises, which corresponded to the placement of the recording devices. "The algorithms can pick up on tiny little clues that confound the actual problems we want to find answers to," she said.

Those who live and work alongside animals, whether they're scientists or not, often think, as a matter of course, that animals can speak with one another, and in depth. Instead of being surprised by the discovery of each "unexpected" animal ability, maybe we should be surprised that humans have such low expectations. Many of us laugh—or shake our heads sadly—when we read that Descartes supposedly threw a cat out of a window to see if it would show fear, as a sort of test for consciousness. (He believed that nonhuman animals were senseless automatons.) Yet many of us would also consider it a wonder that, according to a recent study, elephants seem to have distinct names for one another, which their elephant friends and family use among themselves.

When I started researching this story, I was amazed by each additional avian accomplishment that I learned about, especially in small, ordinary birds. It wasn't only that they communicated this or that to one another but that they were full of concerns—that they were at the center of their own worlds. But shouldn't I have intuited that this was the case all along? I had baselessly assumed that birds had little on their minds. The other day, my daughter and I were walking to her soccer practice, passing by sparrows and also people. "We know almost nothing about birds," she told me. "There's so much we don't even notice." She thought for a moment. "I think they have just as much language as we do, but a lot of it is in their mind. So we don't hear it."

REBECCA GIGGS

The Snake with the Emoji-Patterned Skin

FROM *The New Yorker*

ON A FALL day in Gainesville, Georgia, Justin Kobylka, the forty-two-year-old owner of Kinova Reptiles, was preparing to cut open two clutches of snake eggs. He was hoping to hit upon some valuable, beautiful reptiles. Kobylka is a breeder of designer ball pythons—one-of-a-kind, captive-bred snakes whose skin features colors and patterns not usually found in nature. "I think of myself as an explorer," he told me. Nicking an egg with a pair of surgical scissors, he exposed a live hatchling in its goo. "Even when they haven't yet touched air, you can sometimes see the tongue going," he said, making a flicking gesture with his thumb and fingertip.

We were standing in a six-thousand-square-foot climate-controlled outbuilding that housed some two thousand pythons, which were kept in individual plastic trays slotted into tall metal racks. The space, which cost nearly a million dollars to build and outfit, was immaculate and well lit, with corner-mounted industrial fans and glossy floors. A vague odor of musk and Clorox was all that hinted at the daily chores of snake husbandry.

Ball pythons originated in Africa, and in the wild they are typically dark brown with tan patches and a pale underbelly. Those bred for their appearance, as Kobylka's have been, often have a brighter palette, from soft washes of pastel to candy-colored bursts of near fluorescence. Their patterns, too, have been transformed: a snake might be tricked out with pointillist dots, or a single dramatic stripe, or colors dissolving into one another, as in tie-dye. One captive-bred ball python's splotches and squiggles show up only under a black light. These changes reflect genetic mutations, which breeders call morphs. (The term is also used as shorthand for the snakes themselves.) World of Ball Pythons, a repository of information related to breeding, has catalogued more than seven thousand morphs in the past thirty years—though the actual number likely exceeds that by several thousand. "Evolution can go very fast," the animal-domestication expert and paleobiologist Marcelo Sánchez-Villagra, a professor at the University of Zurich, told me, adding that the variety of "ball pythons may be extreme even among reptiles." Arguably, no other snake, lizard, or turtle has been so sweepingly restyled by human effort.

The animals Kobylka breeds at Kinova are sold to collectors, independent pet stores, and elite breeders who want to replicate, or even improve on, their design. The launch of a new morph is sometimes called a "reveal" or a "drop," echoing the language of luxury-sneaker culture, and there are ball-python internet forums that roil with opinion about which morphs are the hottest, and which ones aren't worth the hype. The most coveted morphs have commanded higher prices than giraffes, lions, and tigers have at auction. "I've had offers of over a hundred thousand dollars on a snake," Kobylka said. "But the way I operate, it's important to keep those snakes for my future work. You actually lose money long-term if you sell the most amazing thing at the time."

Kobylka, who is six feet two, was wearing a gray Lacoste polo, charcoal-colored jeans, and Adidas Sambas; he has dark hair, which he keeps short. In the first clutch, which had nine eggs, he was aiming, he said, for an "orange dream, yellow belly, enchi, leopard, desert ghost, carrying axanthic and clown genes." What sounded like an incantation was a catalogue of desirable mutations. ("Clown," for example, is named for teardrop shapes that show up under the eyes, like the stylized tears of a clown.) As a breeder, Kobylka always has a "goal snake" in mind. "You've done a lot of mental work to imagine it, usually years in advance," he told me, describing a process of zeroing in on specific traits using the known heritability and interaction of genes. "Sometimes your odds are one in two hundred and fifty-six, or one in five hundred and twelve, to make the snake you're thinking about," he said. "The thing that makes it so addicting for me is the fact that there's a large amount of chance involved."

Kobylka has been breeding snakes for more than twenty years and is known as a trendsetter in the field, which is both close-knit and competitive. Courtney Capps, a co-founder of Leviathan Snakes, in South Carolina, observed that buyers are sometimes so proud of owning a Kinova python that, when it comes time to sell its offspring, they'll note in the listing, "Mom was produced by Justin." Kobylka gained an even wider audience in 2016, after he opened up an egg to find a white snake patterned with three orange smiley faces along its body. He had been trying to produce a "dreamsicle"—a white ball python with splotches of tangerine—but most of the circular markings on this snake had two eyes and a grin. Kobylka posted an image on his company's Facebook page, and, when someone suggested in a comment that the photograph had been edited, he made a fifteen-second video that showed him turning the baby python from side to side to display its distinctive motif. The video of the "emoji python" went viral, and the story of the unusual snake was covered by *Esquire, Business Insider*, and the *New York Post*, among other outlets.

On the day I visited Kinova, Kobylka wasn't filming the proceedings, but he sometimes shares egg-cutting videos on Patreon and YouTube. Ball pythons are able to hatch on their own, but such videos, in which a breeder gives a preview of a snakelet's coloration, have garnered a dedicated following. Offering anticipation, disclosure, and irregular reward, they are, in many ways, similar to toy-unboxing sequences. The footage also has elements of the #oddly-satisfying content known as ASMR (autonomous sensory meridian response)—videos of human hands gently manipulating something slimy or soft, for instance, or holding an object ready to burst. It isn't just egg cutting; the entire business of ball-python breeding is extremely online. Breeders deliver pro tips via live stream, develop colorways that will "pop" on Instagram, and often use language borrowed from digital-image editing, promising that a mutation will provide "amazing contrast" or "pixelated sides." Ball-python aficionados can't seem to get enough, finding in morphs a combination of clickbait, dream collectibles, data-driven hobby, and living art. Recently, a video of a ball python with the mottling of an overripe banana next to an actual banana was posted widely on X, while a TikTok video showing a ball python named Gizmo tracing a serpentine line on a tablet computer, as if on an Etch A Sketch, has been viewed approximately five million times. ("He tried to draw himself," one commenter noted.)

It was balmy in the outbuilding. Kobylka modulates the temperature to stagger the pythons' breeding cycles throughout the year (wild females become fertile in response to seasonal cues), but his snakes still seem to intuit the weather outside. A rainstorm can spark mating—Kobylka said that he will sometimes rush to match receptive females to males ahead of a downpour.

The first clutch, I learned, had a parent with a pastel gene, which, in addition to being commonplace, causes the animal's coloring to fade over time. (The standard lifespan of a captive ball python is fifteen to thirty years, though the St. Louis Zoo had one that keepers believe lived to be at least sixty-two.) Pastel unfortunately dominated the brood: most of the snakes would be priced in the low thousands. But, Kobylka said, "every miss is, as probably a gambler would tell you, almost as exciting as a win." A miss, he explained, will still find a home, and can provide useful information about how traits are masked, or about other polygenic effects.

Things went better with the next clutch. Glistening in the first shell was a tiny ball python with three recessive traits—"desert ghost, g-stripe, clown"—and another mutation called "spotnose." The baby snake was the color of straw, with smoky markings down its body, as though it had been repeatedly pinched by hot tongs. "That's everything it could be," Summer Melville, Kinova's business manager, said. The hatchling, a male, would retail for fifteen thousand dollars and be posted on MorphMarket, an e-commerce site for reptiles.

"My wife says I didn't get as excited about our kids being born as when the eggs hatch," Kobylka told me, "but I knew what to expect with our children." (The Kobylkas have five children: two adopted, three biological.) "Actually," he added, "our last son came out with red hair and blue eyes, so he was a double recessive."

As Kobylka went around pulling out trays to show me some of his most valuable full-grown pythons, I was reminded that in nature these creatures are ambush predators. Charles Darwin believed that a fear of snakes is, to some degree, hardwired in us. In *The Expression of the Emotions in Man and Animals* (1872), he recounts putting his face up against the glass enclosure of an adder in an attempt to conquer "the imagination of a danger which had never been experienced." The adder struck at the barrier. Darwin couldn't help but leap backward.

But ball pythons are not venomous, and were named for their tendency to curl up when threatened. The ones I saw tended to huddle in a corner, or contract slowly toward the farthest edge of their trays, which were lined with shredded coconut husk. They couldn't hide their extraordinary appearance, though. I saw a sixty-thousand-dollar python of such stark elegance—bone white and ink—that you could imagine it being unveiled at the Venice Biennale, and a bubble-gum-pink python fit for Barbie. Kobylka admitted to getting quite attached to some of them. Star stickers had been placed on several racks to indicate his favorites, encouraging particular devotion from his staff. In the wild, ball pythons are nocturnal and live mostly underground, often in burrows taken over from rodent prey. They are not very social, though infants may stick together for a short time after hatching. Was Kobylka's affection, then, one-sided? "Ball pythons don't seem to mind being held, but they don't seem as curious as some other species," he said.

The team at Kinova would soon box up dozens of the company's finest specimens and drive them to Tinley Park, Illinois, for the North American Reptile Breeders Conference (NARBC), one of the most anticipated reptile expos in the nation ("the

Mecca of the ball-python market," as one breeder called it). Kobylka, who owns several Porsches, including a 2016 Cayman GT4 Clubsport that he races, compared the NARBC to a premium car show, offering a window onto the future of the industry. A draw for attendees this year would be the opportunity to meet Emily Roberts, the star of Snake Discovery, a YouTube channel with more than three million subscribers. Kobylka himself was teasing the existence of a baby "sunset combo" on YouTube. "We finally hit something really epic," he announces in the video.

Before leaving Kinova, I asked if I could hold a ball python. Kobylka selected a small lemon-yellow snake and placed its rolled-up body on my open palm. I had expected something cool to the touch, but the snake was warm, the temperature of its enclosure. When I shut my eyes, the impression of it on my hand seemed remarkably faint. "They tend to just sit still, and they're handleable," Kobylka said. "They're just so packaged."

The ball python is known to zoologists as *Python regius*, or "royal python." Cleopatra is rumored to have worn one as a bracelet, but the story is almost certainly apocryphal—ball pythons have never been native to Egypt. Today, ball pythons live in western Africa and parts of central and eastern Africa, from Senegal to the borderlands of South Sudan and Uganda. About three feet long, they can be found on the margins of rainforests and in the woods, but they have also adapted to managed environments: timber plantations, agricultural fields, trash heaps.

Snakes have evolved without major transformation for more than a hundred million years. Ball-python breeding and collecting is a relatively recent phenomenon. The high end of the American reptile market was long monopolized by large, heavy-bodied snakes, like boas and reticulated pythons. In the early twentieth century, dealers mainly sold their wares to film studios and zoos. By the mid-1960s, improved habitat construction in zoos led to snowballing competition for hard-to-collect species—Angolan pythons plucked from war zones along the Namibia-Angola border, iridescent Boelen's pythons caught on the mountainsides of Papua New Guinea. The consumer market for pet reptiles was sluggish by comparison; through the 1980s, wildlife traders viewed parrots as more profitable. The investigative journalist Bryan Christy has described reptiles as having been "the Bic lighters of the pet industry: cheap, disposable point-of-sale pets."

In the early 1990s, though, household reptiles began to get a reputational makeover. Children raised on *Teenage Mutant Ninja Turtles* and *Jurassic Park* reimagined scaly pets as characterful and intriguing. Retailers started to see an uptick in iguana sales. New Caledonian crested geckos, believed extinct until 1994 and jeopardized today by wildfires and invasive predators, became well established in captivity. Snakes were pitched to prospective buyers as perfect for cramped urban residences: undemanding, hypoallergenic, and needing to be fed only once a week. Ball pythons—which were abundant in their natural habitats and, being compact and docile, highly transportable—were soon arriving in the U.S. by the crateful, tucked into sacks and pillowcases.

Between 1989 and 1999, exports of ball pythons from West Africa to the United States tripled. In *The Ultimate Ball Python,* an encyclopedic volume on morphs helmed by the breeder Kevin McCurley, one early broker described them as a "junk" species. In pet stores across the U.S., imported brown-and-tan ball pythons sold for around thirty dollars, discounted for being less alluring than other tropical reptiles. They were quintessential starter pets, and when he was in his mid-twenties McCurley, now an exuberant figure in the snake world, owned two: Eek and Meek. Speaking to me from what he called a "venomous room" in his breeding complex in New Hampshire, McCurley said that one day he had a vision of what ball pythons could be: "I looked at the vehicle of the ball python, and I said, 'This is the ideal snake. But it needs a totally different paint job.'"

In 1989, an Oklahoma-based breeder named Bob Clark received a tip about a single albino ball python found in Africa. "I got a letter from a friend in The Hague, Netherlands, about a dealer in Ghana, who had the animal," he told me. For a few years, Clark had been successfully cultivating albino Burmese pythons, costly novelties that can reach almost twenty feet, which he raised on rats, rabbits, and piglets. But in the ball python's small size he saw an opportunity. On the strength of a photograph sent to him in the mail, he bought the albino for seven thousand dollars, a price that he said "seemed a little crazy" at the time.

It took Clark several years of line-breeding—mating snakes to their forebears, littermates, or descendants—to produce a second albino ball python, and more followed. (Such ball pythons aren't pure white: you might get a snake with carrot-orange daubs or pale-yellow streaks.) He began selling the hatchlings at seventy-five hundred dollars apiece, and a wait list quickly formed. Collectors liked them, but his main type of client, he found, was the aspiring breeder. "Everybody wants nice, beautiful, expensive snakes that are rare," he said. "One way they justify that to themselves, and their spouse, is to say, 'This could be a moneymaker.'" Then, in 1994, Clark's facilities were burgled. The thief got away with his founding albino male as well as females that were heterozygous for the trait. Clark retained a large enough colony to continue, but he began hearing rumors of other albino ball pythons: his supply was no longer exclusive. (The thief was eventually caught and ordered to pay a civil judgment of $2.5 million.)

Soon another ball-python type, the piebald, or pied, which features mottled brown spots on an ivory body and is considered scarce in the wild, became popular: in 1997, a pied ball python could sell for thirty thousand dollars. McCurley, who had been breeding reptiles while working a day job as an electronic technician, couldn't afford an albino, let alone a pied. Instead, he started buying imports with minor irregularities, and mating them to determine whether a specific quirk could be passed on (a process known to breeders as "proving out"). Most anomalies were discreet—a bit of speckling, a squiggle along a spine. Matthew Lerer, who used to sort reptile shipments in South Florida, noted in *The Ultimate Ball Python* that McCurley would study the snakes' markings for hours, "like he was a gemologist inspecting the Hope Diamond." Mike Wilbanks, a snake breeder from Oklahoma, told me, of the years that followed, "Some of the morphs turned into gold mines. Some turned out to be just a dry, empty hole." It is not possible to trademark a morph, but breeders came to view the particular designs they were working toward as commercially sensitive information. The first to produce a morph, and name it, gained celebrity.

Many breeders believe that ball pythons' history of living primarily belowground has preserved an array of mutations related to appearance. The thinking is that eye-catching snakes living aboveground are more visible to predators, making them more liable to be picked off before their genes get passed on. (Ball pythons have poor color vision, and their markings, unlike those of many lizard species, are not thought to play a role in courtship.) By the early 2000s, middlemen in Ghana, Togo, and Benin had learned that American buyers were willing to spend top dollar for "odd balls"—snakes that diverged from the wild type in even minor ways. What had been an amateur pursuit was fast becoming an industry. Ball-python exports from West Africa peaked at around two hundred and fifty thousand a year in 2005, and began to decline, as domestic breeding replaced mass importation. American captive-bred ball pythons seemed to better express buyers' notion of the exotic. The wild type had begun to be seen, in McCurley's words, as "garden variety."

Perhaps not unrelatedly, snakes—and snakeskin—were having a moment in the broader culture as emblems of opulence and transgression. At the 2001 MTV Video Music Awards, Britney Spears sashayed across the stage with a seven-foot-long Burmese python slung around her shoulders. Gleaming in the windows of high-end fashion boutiques were python-skin footwear and clothes, often dyed traffic-light green, neon yellow, or electric blue, from Yves Saint Laurent, Jimmy Choo, and Chanel. ("Eve's Revenge, the Python's Sorrow," one headline in the *Times* ran.) Ball pythons were too small to be profitable for the skin trade, but the Zeitgeist's embrace of surreally hued scales conferred an aura of glamour on collecting and breeding them.

Kobylka had an itinerant childhood. His mother—who raised him and his younger brother on her own—sold handicrafts and moved the family to Nebraska, Oklahoma, Colorado, Tennessee, and Arizona. Wherever he was, Kobylka spent hours outside. "I always felt there's a world within a spot," he said. "Lizards and turtles and frogs, centipedes and salamanders—creatures most people would walk right past and never see." His mother encouraged Kobylka's pursuits but insisted that no animal be brought home. "That was the rule with my family: what's wild is wild."

When Kobylka was a teenager, he began attending a small religious boarding school in Oklahoma, where he became fascinated by scarlet king snakes, an elusive, tricolored species. After class, he would go searching for them, turning over logs and debris. He didn't find one, but he did catch a rattlesnake, which he kept in a homemade cage in his dorm room until he himself was caught. "I met my wife at the school, and it was her mom who called and tattled on me," Kobylka said with a laugh. The school did offer him an empty room to accommodate the rest of his collection, which by then included White's tree frogs, box turtles, and a red-tailed boa he had bought at a pet store.

At nineteen, Kobylka spent a gap year in Benin, where ball pythons pocket the landscape. While there, he visited a Vodun temple and saw ball pythons, which are regarded as sacred, roaming freely. "I have pictures of me from that time, holding ball pythons, having no concept of: this is going to be my whole life," he said. What captivated him then were chameleons—reptiles that change color for camouflage or to indicate excitation, rivalry, or submission. He gave local children pocket change for any that they could catch, and placed the animals on the boughs of a tree outside his lodgings. He would sometimes climb the tree and be surrounded by the creatures, the shade tick-ticking with eyes.

Kobylka attended Southern Adventist University, in Tennessee, where he majored in communications and began keeping king snakes—which are banded and slender—caged in his dorm bathroom; he soon had fourteen of them. He was reintroduced to ball pythons through the website of a Maryland-based breeder named Ralph Davis, who kept pieds. "He had this rock-star personality," Kobylka told me. "Ralph's site was the only place you could see all these mutations and get a picture of what was possible. Everyone else was stuck in the Stone Age." On weekends, Kobylka would visit his uncle, a physician in Georgia. "I was just talking about snakes constantly," Kobylka said. "I drove him crazy."

One day, his uncle offered to invest in a pair of pastel ball pythons, then top-tier morphs, and to split the profits if they produced salable offspring. They earned enough on the deal that Kobylka was able to persuade his uncle to buy a pied, which he went to fetch from a dealer in Florida. "I still remember driving home with that animal," Kobylka told me. "I would stop by the side of the road to look at it every hour. And I just—" He exhaled, seemingly at a loss for words. "Amazement," he said after a beat. (These days, ball pythons are sent between dealers and collectors using overnight mail carriers such as FedEx.) His uncle attached one condition to the purchase: the snake was too valuable to be left in Kobylka's care, so it would live in a tank next to his uncle's bed.

The early 2000s were a good time to get into the business. McCurley had received a six-figure bid for three golden snakelets with webbing patterns, called "spider" morphs. The breeder Mike Wilbanks sold a "black-eyed Lucy"—a leucistic ball python—to a Belgian collector for two hundred thousand dollars. People were taking on debt to finance their ball-python purchases. "I had second mortgages on my house so I could have a hundred thousand dollars ready to go if that next new thing came out," Wilbanks told me. "It was a big race."

Like a new gadget, a morph might be faddish and expensive at first, but as it was sold widely its value would slide to an entry-level price point. Those rare ball-python traits first discovered in the wild and now known as base morphs had followed this trajectory. "They all became accessible," Kobylka said.

He recalled thinking, "Wait a minute, there's infinite possibilities if we just stack genes together." He told me, "That's where I got my jump on the industry." In 2003, he launched his business, aiming for "combos," mutations layered together, in order to produce singular dazzlers that could appeal to connoisseur breeders in the U.S. and internationally rather than to big-box pet stores. It would be a slow-release undertaking. "I wanted to imagine a morph combo in the future, and create it ten years later—that was what I was all about," he said.

These days, Kinova's yield is around fifteen hundred pythons a year, and Kobylka is seen as unrivalled in his artistry with genetic mutations, a breeder at the frontiers of the form. "You will never ever be able to catch Justin Kobylka," Antoine Hood, of High Desert Pythons, in North Carolina, told me. "How can I parallel him? That's a better endeavor." Brittney Gobble, who breeds ball pythons in Tennessee, called Kobylka "a savant." "He's able to create a 3-D model, combining five or six mutations in his mind. And then eventually it gets made, and it looks like how he described it," Gobble said. "It's truly just insane."

On a brisk October day, I went to the North American Reptile Breeders Conference in Tinley Park, where venders from around the country had set up booths. With most transactions taking place online, expos are seen as an opportunity to launch what Kobylka called "really cutting-edge things." Many breeders showed their ball pythons in clear acrylic boxes on logo-printed paper; photographs of them would be automatically branded.

Kobylka had walked around the venue before it opened to the public, taking note of the competition, but now he couldn't get more than a few steps away from the Kinova booth without being accosted by neophyte breeders and reptile enthusiasts. In the melee, I saw a buyer lightly rap the top of one of Kinova's display cases and announce, "That's the snake—that's the showstopper." It was the "sunset combo" morph, which had the bittersweet-orange sheen of heirloom glassware. Kobylka held court, fielding queries and dispensing advice. "Justin would take this off my hands in a second, if I let him," a breeder who had brought a morph called Cyborg said. Emily Roberts's fans had dressed in pink, and together they formed a shuffling, spangled queue. A woman in a sleek blazer was deftly handling a lustrous slate-blue snake, changing her grip the way a rappeler belays a top rope, hand over hand, as the snake cascaded without progress in the direction of the floor. "They can have quite a spicy temperament," she declared. "They're not for beginners." At the booth for Best Dressed Balls—an Iowa-based venture run by a breeder named Troy Schroeder—a girl of nine patted the box in front of her, fixated on the creature inside. "Tell everyone he's unlovable, Troy," she pleaded, hoping to save up enough to bring it home.

An estimated six million households in the U.S. include at least

one reptile. Millennials make up the largest group of reptile owners, but snakes, lizards, and turtles have become increasingly popular with Generation Z. "One of our concerns is that technology will
take kids away from this world," a breeder observed. "Why would a
kid today want to peer at a snake through glass, when they can put
a VR headset on and play with dinosaurs?" As much as the ball python seems to have been pulled into the technological infrastructure of the twenty-first century—featured in live streams, traded
via MorphMarket—snake ownership was frequently portrayed at
the expo as an antidote to the anomie of feeling ourselves to be
part of a big machine. A 2015 poll of readers by *Scientific American
Mind* found that snake owners were more likely than other pet
owners to describe their animals as "part of the family."

Some reptile owners clearly felt that more was more. Bob
Clark was in the crowd, buzzing from the recent sale of five
"retics" (reticulated pythons) to a customer in the Middle East,
for half a million dollars. These snakes were so big that, once
crated, a forklift had been needed to move them. Among breeders, the matter of snake size could be divisive. "One percent of
snake keepers are up to taking care of a snake that large, and
not the other ninety-nine," Kobylka said. He expressed unease
at the rise of social-media accounts that sensationalize living
with gigantic snakes, misrepresenting snake keeping as an extreme sport rather than a serious responsibility.

But the majority of collectors were there for the ball pythons.
Although *Python regius* is not endangered in Africa, the International Union for Conservation of Nature (IUCN) has designated
it as near threatened, and the NARBC prohibits animals that ar-
en't captive-bred or "quality farm-raised" in the U.S. As I wandered through aisle upon aisle of ball pythons, I wondered if the
line between the wild type and the captive-bred could be so easily
demarcated. Rob Rausch, who is in charge of juvenile animals
at Kinova, told me that he thought ball-python morphs took the
pressure off wild-snake populations by satisfying people's long-
ing for exotic-looking reptiles. "You can go to the wild, and you
can get a normal ball python. Or you can come here, and"—he
made an openhanded gesture, as of a shopkeeper displaying his
wares—"What color do you like? What pattern do you like?"

At the NARBC, combos that were darker in tone reigned. I saw rows of axanthics, which are monochrome, in ash white, pewter, and black. There was a "hurricane" combo, with lightning slashes of electric yellow along its chocolate-brown sides. Then there were the ball pythons that hadn't yet been conjured. Kobylka told me that he was hoping to make "a truly zebra-looking animal, with black and white stripes." He added, "We visualize that, but the difficulty is that'd be a quadruple recessive. That's many years out." I had already seen, at the expo, designer "leopard" morphs, with the coloration of a spotted big cat, and there were also two much-talked-about "gorilla" combos, which were dark with tortoiseshell ripples: as on fashion runways, there was a constant conflation of wildness with luxury.

Ball pythons have come to be seen as unnatural, hothouse creatures. At the conference, I frequently heard them described as "pet rocks"—that is to say, inert and highly collectible. "A snake is a snake," Clark said. "You can't make a dog out of it." I appreciated the fact that many breeders seemed to resist anthropomorphizing their stock. But speaking of ball pythons as "pet rocks" seemed to ignore their fundamental creatureliness. "I've had people who have had five-thousand-dollar, ten-thousand-dollar snakes, who said they didn't want to pay seventy-five dollars for an exam or treatment for that animal," Mark Mitchell, a professor of zoological medicine in Louisiana, said. "People are much more apt to want to take care of an animal they view affectionately than those they consider commodities."

Yet evolving into eye candy for humans has meant that designer ball pythons, when viewed at the species level, enjoy some of the evolutionary advantages of domestic animals, including wide dispersal. "There isn't a single endangered species of domestic animal," Marcelo Sánchez-Villagra told me. "Many have worldwide distribution." With natural habitats disappearing all the time, finding a way to shelter within anthropic culture might be a good strategy. "What is a better ticket to survival than being beautiful and rare?" Clark said. "Those traits are going to be multiplied in future generations because that's what people like, not because that's what kept you from getting eaten."

The evolutionary trade-offs borne by individual snakes in

captivity, however, have, in some instances, been dire. Breeders once championed their field as a salve against the cruelties of shipping wild-foraged snakes, which can be at risk for dehydration, parasites, and increased disease transmission, but morphs aren't always better off. Over time, breeders have discovered that several sought-after traits and specific gene-crossings also produce physical irregularities. Duckbill, in which a snake's rostrum is upturned and flattened, is a benign deformity, but it is said that some morphs, such as the "caramel albino," have a higher chance of producing young with spine kinks, a condition that can prevent them from moving sinuously, or can fatally obstruct digestion. "If we hatch a python with small eyes, we won't sell it as a breeder," Courtney Capps told me. "We sell those as pets only, because I don't want the genetics to be passed on. Potentially, at some point, small eyes turn into no eyes." Another condition that can be distressing is "wobblehead," a tic that likely betrays neurological issues. Online, breeders have counselled against pairing morphs that are known to result in impairments. ("We want mutations that are just skin-deep," Kobylka told me. "We don't want the animal to be at all changed in any way that would hurt its ability to survive.") But some unusual malformations, unrelated to breeding morphs, can be profitable windfalls: early in the new year, Clark sold a two-headed ball python for a hundred thousand dollars. "Both heads eat," he assured me, when I inquired after its health.

At Kinova, I had asked Kobylka which was the rarest snake in the room, expecting to be shown something supremely expensive. As it turned out, though, the rarest snake was an endling, the last of its kind, and wouldn't be sold. He retrieved it from a tray. The morph was called "desert," a snake the color of burnt butter with a toothlike pattern. It was a variety that "everybody loved when it first came out," he said, "but, if you try to breed them, the eggs will get stuck and they'll die."

Unsurprisingly, many snake experts are skeptical of the whole morph-making enterprise. "It's *The Island of Doctor Moreau*," the British herpetologist Neil D'Cruze told me. D'Cruze is the head of research at World Animal Protection International, and the senior co-author of "Snakes and Ladders," a scientific paper on the ball-python trade, published in 2020. "The speed-breeding, the genetic manipulation, it's being pushed out of the desire to create a new product. Not to help the snake cope better in captivity—to be a better pet for whoever owns it. Are these animals part of a genuine conservation program to help save the species? No."

Some are concerned about the limits of our ability to envision what snakes need, and to act in their best interests. Eben Kirksey, an Oxford professor of anthropology who has written about python-breeding communities in the U.S., believes that seeing "past the dollar value of a snake with particularly colorful skin" would mean offering more to the snakes than racking trays. Breeders "talk about burrowing," Kirksey said, "but the enclosures I've seen, they're not like actual burrows. These are life-support technologies that people are cobbling together out of plastic, out of machines." Were the snakes okay with all of this? "There are a lot of animals that, unfortunately for them, tolerate captivity well," D'Cruze told me. "But suffering isn't always overt. Suffering can be under the hood, invisible." I had read that it was possible to gauge a python's stress by measuring its blood cortisol, but as I walked around the expo I found myself troubled by the question of what thriving or discomfort looked like in a snake. Could a python raised in a tray, fed, kept warm and watered, and bred be said to live a full life?

In the meantime, the business of breeding rolls ever forward. When ball pythons were first becoming investment pieces, inevitably there were scams. One individual purported to have the world's first entirely red ball python, and sold the python's offspring—which were all black—to several U.S. concerns. As clutches laid by those snakes failed to contain any crimson hatchlings, vexed breeders agreed that they'd been conned. New technology promises to change all of that. A topic of fervent conversation at the expo was shed-testing, a kind of 23andMe for snakes, as Kobylka put it. "That's where our industry is headed," he said. The testing, which requires the tiniest scrap of molted snakeskin, offers designers a more rigorous way to verify traits. Brittney Gobble told me she had heard that artificial insemination might soon be available for ball pythons, which would expand the field exponentially. I imagined a world where creating new morphs would be a matter of transporting little vials of snake sperm, not the snakes themselves.

In one of our earliest conversations, Kobylka had said that he wanted "to make something that is genuinely beautiful to an average person. That's my criteria—if the person on the street, who doesn't like snakes, stops and says, like, 'Whoa, that's a snake?! I didn't know a snake could look like that.'" But it was also possible to go too far. A scaleless morph, for example, has bald, matte skin, similar to a sphynx cat or a furless guinea pig. Many will recognize it as "the pinnacle of 'unnatural,'" the veterinarian H. Kitt Hollister wrote in *The Ultimate Ball Python*, because it is "delicate, and seemingly unable to survive without human intervention." The breeder responsible for the morph, Brian Barczyk, who died earlier this year, of pancreatic cancer, marketed it as "smooth and soft," a completely different texture to touch. Kobylka told me that when he finally got to hold one he was perturbed. The scaleless snake seemed to break the boundaries of what a snake is. "They feel like human skin," he said, shuddering at the memory.

BEN GOLDFARB

Relearning the Language of the Land

FROM *Smithsonian*

THE TWENTY-EIGHTH QUESTION of the hardest exam I'd taken since college was a smudged footprint pressed lightly into damp sand. The track possessed four teardrop-shaped toes and a vaguely trapezoidal heel pad; one toe, third from the left, jutted above the others, like a human's middle finger. I knelt closer, scanning for the telltale pinpricks of claws, and saw none. That suggested feline: cats, unlike dogs, have retractable claws, and they tend to walk without their nails extended. In my notebook, I wrote, tentatively, "BOBCAT."

The hypothetical bobcat had been wandering a sandy flood-plain in the California desert, where I found myself taking a wildlife tracking test one April afternoon. The evaluation was administered by Tracker Certification, the North American wing of CyberTracker Conservation, a South African nonprofit that has conducted tracking exams for thirty years. Around me, other students were engaged in their own examinations—peering at a bone-filled lump (great horned owl pellet), inspecting snipped willow stems (woodrat chew), contemplating stick-like prints at the creek's edge (thirsty scrub jay tracks). The vibe was library-like, studious and hushed, as we attempted to read the land's open book.

We were participating in an art as ancient as hominids ourselves. Tracking, by allowing humans to more effectively pursue game, drove us into hunting groups, grew our brains, compelled us to adopt language. In his 1995 book, *The Demon-Haunted World*, Carl Sagan posited that tracking shaped our evolution: "Those with a scientific bent, those able patiently to observe . . . acquire more food . . . they and their hereditary lines prosper," he wrote. Tracking helped to transform a small, furtive ape into a global force.

No longer does our survival depend upon our ability to stalk a springbok. *Homo technologicus* is more attuned to screens than to scats; the trails we follow are paved highways rather than pawprints. Even the field of wildlife biology has become reliant on technology. Scientists use satellite collars to monitor caribou from their desks; drones hover over penguins in Antarctica; motion-activated cameras snap photos of every creature that crosses their infrared beams. Today a herpetologist can identify each frog, toad, and salamander in a creek by sifting through snippets of DNA in the merest vial of water. Crouching over skunk prints and jackrabbit pellets feels analog by contrast, even anachronistic.

Yet old-school tracking—a cheap, noninvasive method capable of providing astonishing quantities of data—is experiencing a revival. These days biologists are examining tracks for many purposes, from forestalling wildlife conflicts to averting roadkill. In Wisconsin, trackers are following wolves to prevent them from running afoul of livestock and humans; in Washington State, they're observing faunal footprints returning to river valleys after dam removal. Biologists have trailed the antelope-like saola through Southeast Asia's mountains in hopes of capturing and breeding it, and stalked lynx and wolverines across Montana to understand their abundance and distribution. Many tracking-based studies make use of data collected by volunteers, who, with training, can follow animals as skillfully as academic scientists. "It can be this very accessible, democratic way of gaining information," David Moskowitz, a naturalist, photographer, and expert tracker, told me.

In the past two decades, Tracker Certification has conducted nearly 700 formal field evaluations, during which it accredited more than 2,300 individual students. They make up an eclectic array of people. The participants in my workshop included photographers, teachers, and hunters; there were biologists, yes, but also chemical engineers and real estate brokers who spent their weekends volunteering on mammal surveys. At the morning's outset, we'd gathered in a sandy wash and, squinting into the rising sun, explained our sundry motivations. One student said he longed to "read the little letters that the world is writing you"; another declared that he "didn't want to feel like a tourist" in the natural world. "You're all helping to revive real field skills and natural history skills across the planet, in a way that is desperately needed," Casey McFarland, Tracker Certification's ebullient executive director, had declared. Then he'd released us—to scrutinize spoor, to pore over prints, to track.

For nearly the entirety of human existence, our species regularly performed feats of tracking that, from our modern vantage, resemble magic. So Elizabeth Marshall Thomas learned one day in the 1950s, when she set out with three Ju/'hoansi, hunter-gatherers in Namibia's Kalahari Desert, on the trail of a hyena. Thomas's parents, ethnographers and adventurers, had moved their family to the Kalahari when their daughter was nineteen, and the Ju/'hoansi's practical brilliance instantly awed Thomas. The Ju/'hoansi shot wildebeest and other game with poisoned arrowheads, then tailed their dying quarry for many miles. This required not only distinguishing tracks at the species level, but also recognizing individual animals: if a herd of kudus split up into several bands, the Ju/'hoansi had to remain on the trail of the particular kudu they'd shot. It was a skill "that must be seen to be appreciated, especially because none of the tracks are clear footprints," wrote Thomas, in her 2006 book, *The Old Way*. "Mostly they are dents in the sand among many other scuffed dents made by the other kudus."

Yet even Thomas couldn't believe how her companions followed the hyena across an expanse of bare rock. "They were not simply following the line of travel, because out on the rock, the route of the hyena made a curve of about one hundred degrees," she wrote. There were no footprints, no drops of blood, no bent grasses. Still, when the party reached the sand beyond the plateau, there resumed the hyena's tracks, exactly where the men expected. "How they did it I have no idea."

North America's Indigenous peoples, of course, shared an equal intimacy with tracks. "It's culture for me—it's in my religion and my people's history," Ahíga Snyder, a Diné, or Navajo, wildlife researcher who co-runs Pathways for Wildlife, a research group in California, told me. Tracks featured prominently in the Diné animal stories that Snyder's grandfather told him. Take Black Bear, whose front paws have straight, evenly sized toes that can sometimes make it hard to tell a right foot track from a left—because, per Diné legend, the bear woke up late on the day that the Great Spirit assigned each animal its tracks and, in his haste, slipped his moccasins onto the wrong feet. Or consider Deer, whose hooves formed an arrow said to point toward prosperity, appropriate for such an important game animal. Understand tracks, Snyder said, and "the whole world opens up differently."

Scientists held tracking in high esteem well into the twentieth century. In a 1936 paper titled "Following Fox Trails," the biologist Adolph Murie described spending months shadowing red foxes in Michigan. (Sifting through their droppings, Murie reported "many rabbit bones and pieces of fur, the skull and a piece of the hide of a muskrat, the rear half of a fox squirrel, the posterior part of a skull, scapula, and femur of a lamb, part of an adult deer sternum, and some blue jay feathers.") As technology improved, however, biologists adopted tools like satellite collars and motion-activated cameras, which allowed researchers to glean vital data across vast areas with less labor. Tracking fell out of favor.

In some ways, tracking's revival began in the 1980s, when a South African physics student named Louis Liebenberg abandoned his studies, bought a Land Cruiser, and drove into the Kalahari Desert to commune with the San, a group of hunter-gatherers that includes the Ju/'hoansi. Liebenberg had always loved to track; while serving in the military, he'd occasionally abandon his post to sketch animal footprints. Liebenberg spent years with the San, watching them track and nearly dying of heatstroke during an antelope hunt himself. Tracking was not antithetical to science, he concluded. It was science. Following animals inherently required observation and inference; what was pursuing a trail if not testing a hypothesis? Just as physicists deduced the existence of subatomic particles from the movements of visible matter, San hunters reconstructed the behaviors of invisible animals from tracks. It was conceivable, Liebenberg wrote, that "the creative scientific imagination had its origin in the evolution of the art of tracking."

That art, however, was at risk of disappearing. Many San neither wrote nor read; meanwhile, ranch fences across the Kalahari had curtailed wildlife migrations and threatened hunter-gatherer life ways. Around campfires in the early 1990s, Liebenberg and a San named !Nate began to discuss how to simultaneously preserve traditional knowledge and provide Indigenous trackers a credential they could use to secure jobs as ecotourism guides and park rangers. In 1994, Liebenberg began conducting training workshops and accreditation tests for San trackers. He and others also developed an elaborate pictorial system, first operated on a PalmPilot and later on smartphones, that allowed trackers to record and share their observations. They called the system, released in 1996, CyberTracker.

Together, the CyberTracker software and Liebenberg's certification process proved their worth. Non-literate San co-authored peer-reviewed papers on topics such as black rhino behavior and found work defending animals from poachers. In 2002, hoping to expand his system to the United States, Liebenberg attended a wolf tracking workshop in Idaho, where he met Mark Elbroch, a young wildlife biologist who'd cut his teeth studying mountain lions in Wyoming. Elbroch spent portions of the next three years in the Kalahari, trailing animals from lions to porcupines with Liebenberg and his San colleagues and, between trips, applying his newfound knowledge to black bear and cougar tracks at home near Santa Barbara. Elbroch eventually aced his exams in the Kalahari and Kruger National Park, and he and Liebenberg set about importing CyberTracker's protocols to the United States. In 2004, they conducted an evaluation in the California desert and, the next year, Elbroch led one for staff at the Texas Parks and Wildlife Department. "At first they were confused and frustrated: 'What the heck are we looking at?'" Elbroch recalled. "By midday, everyone was having the time of their lives." Cyber-Tracker's standards had been developed on zebras and African lions; they would be honed on mule deer and cougar.

These days Elbroch lives on the Olympic Peninsula, the wedge of temperate rainforest that juts from western Washington like a thumb. One autumn morning, I set off into the Olympic woods with Elbroch and Kim Sager-Fradkin, a wildlife program manager with the Lower Elwha Klallam Tribe, to reconstruct a day in the life of a mountain lion. The cat was a two-year-old male, named Orion, whom scientists had outfitted with a satellite collar under the auspices of a long-term study called the Olympic Cougar Project. Mountain lions are generally peripatetic, but, according to Orion's collar, he'd recently lingered in one area for more than a day—a hint that he'd made a kill and hunkered down to eat. Now Elbroch hoped to find Orion's meal and piece together the circumstances surrounding his feast.

The three of us tramped through stands of alder and shafts of sunlight. "Bobcat scrape," said Elbroch, who sported a gray-flecked beard and a T-shirt bearing his own detailed illustration of a cougar skull. He nodded to a barely perceptible patch where a cat had cleared the leaf litter and urinated to mark its territory. Moments later, we came upon another, larger scrape. "Notice the size difference? This is probably our boy." He picked up a broken fern frond, browned and curled at its edges. "Following animals is all about color and pattern," he whispered.

Beyond the scrape ran a faint swath of exposed ground where Orion had hauled his prize. "Whatever's at the end of the rainbow, this is where it starts." He turned to me. "All right, now you're loose. Lead on."

I blanched. Mark Elbroch asking me to trail a cougar was like Jimmy Page suggesting I play a few chords. "They almost always drag downhill," he said. I stumbled down a steep slope, clambering over moss-garbed deadfall in pursuit of an apparent dragline. At the bottom, I paused to get my bearings. Nary a scrape to be seen; the ferns looked as fresh as though they'd sprouted that morning.

"We're up here," Elbroch called from a ridge.

"I thought they dragged downhill," I complained. He smiled and shrugged.

I rejoined Elbroch and Sager-Fradkin as they traipsed along, pointing out what trackers call "sign," or any animal trace besides a footprint: the exposed root where the carcass had rubbed off moss, the matted earth where Orion had curled up to digest. Plate-size globs of bear scat lay everywhere, and the bruins' beds cratered the forest floor. "Bears are big, lumbering beasts that leave a lot of sign," Sager-Fradkin said. "They're not delicate and neat like cougars are."

She and Elbroch were looking for the cache, the trove where Orion had hidden his leftovers. Elbroch paused before an unassuming smudge of dirt, perhaps bloodstained, and began to root with a stick. The odor of rotting meat wafted up. He plunged his hand into the soil and, with the flourish of a magician revealing the ace of spades, held up the skull of a mountain beaver—an odd rodent, unrelated to true beavers, whose burrows pockmarked the hillsides. A few inches deeper and he unearthed an off-white fragment of zygomatic arch, a bone that once resided in the head of a small deer. A narrative cohered: Orion had killed a fawn, lugged it here to consume and stash, and, between helpings, captured a passing mountain beaver, as though nabbing a bacon-wrapped scallop off an hors d'oeuvres tray. Before Orion could disinter his fawn, though, he'd been run off by a bear. There was no way to confirm the story's veracity, but it felt entirely plausible.

This exercise wasn't only an enthralling party trick—it also had profound value for the Olympic Cougar Project. While satellite collars could tell Elbroch and Sager-Fradkin where cougars approached I-5 and other highways, thus guiding the location of future wildlife passages, only tracking could reveal what they ate. Figuring out how many deer and elk the peninsula's cougars killed, for example, could help tribes determine hunting quotas, thus keeping game on the landscape for felines and Native people alike. Studying the dietary preferences of males like Orion, who often run afoul of humans while searching for territories, is especially vital. "The naive ones don't really know where to go and how to avoid people," Sager-Fradkin said. Instead, they occasionally develop the unfortunate habits of attacking goats, calves, and other livestock, and following small prey like raccoons into neighborhoods until they learn to hunt deer. Understanding the diets of these rambunctious cats may help prevent conflicts between the carnivores and people, and protect the species upon which dispersing cougars rely.

"We're redefining these animals to hopefully learn how to live with them," Elbroch said as we crouched in the duff, examining bone shards. There are still questions that can only be answered by digging in the dirt.

After my experience trailing Orion, I resolved to try tracking myself. I bought one of Elbroch's guidebooks and began canvassing my home landscape in Colorado. I marveled at a twisted mink scat packed with a snake's tiny skeleton and the pointillist artwork inscribed in aspen bark by a black bear's claws. Yet my callowness left me unfulfilled. Consulting Elbroch's books might lead me to *believe* that a pawprint had been left by a red fox rather than a gray, but I couldn't be certain. I craved validation.

That was how I ended up in California, scrutinizing bobcat prints in hopes of passing a tracking test. The exam's format was simple, its content challenging. Beforehand, our evaluators—McFarland, Tracker Certification's executive director, and Marcus Reynerson, a senior instructor at a wilderness school—had scoured the desert for animal tracks, scat, gnawed twigs, and other oddities. They planted an orange flag at each impression; our charge was to figure out what was responsible for the various marks. (Some questions were complex multiparters that required us to identify not only what animal had left a given track, but also its pace and the responsible foot: left front, right hind, and so on.) Once we'd devised our answers—which, in my case, could charitably be described as semi-educated guesses—we whispered them, or showed them in writing, to clipboard-wielding assistants.

From a distance, the desert appeared a barren expanse of rock and sand, but up close it throbbed with life. We were asked to explain the provenance of a bone shard (a coyote's scapula) and a silk-lined hole (a tarantula's burrow) and a white crystalline splatter (wonderfully, calcium from the urine of a desert cottontail). Most of all, we had to know tracks. A raven's inner toe was tucked close to its middle, whereas a jay's toes were clustered together. Two long paddle-shaped marks, punctuated by a pair of small circles, were the hind and front feet of a resting jackrabbit.

Each time we finished answering a batch of questions, we gathered to debrief. I'd been dreading this part, which promised to expose my ignorance, but I needn't have worried: the vibe was less oral examination and more Socratic dialogue. How do you tell a domestic dog's claws from a coyote's? (Among other differences, dogs leave blunter marks, since their nails are clipped or filed down on pavement.) How do we know this turd came from a toad? (It's gleaming with insect exoskeletons.) McFarland and Reynerson, good-natured and nonjudgmental, greeted even incorrect answers with an exclamation of "Fantastic!" Wrongness was an opportunity to learn.

This point was reinforced when we were asked to identify the source of chew marks on a prickly pear. "I thought it was a mule deer," one student said. (I'd guessed bighorn sheep, which often eat cactus in the desert.) "Excellent!" McFarland enthused. Then he pointed out what was missing from the scene: the spines of the cactus. A deer or sheep would have nibbled the prickly pear's flesh around the needles, which would then have fallen to the ground; their absence suggested that some creature had carried them off. This, then, was the toothwork of a woodrat, which had borne away the spines to armor its nest. "They're getting a food source, but they're also getting protective equipment," McFarland said. Even a diminutive rodent wrote upon the earth.

Sometimes, as we perused tracks and sign, I was appalled by my ignorance: How could I have confused that rabbit pellet for a squirrel's? Other times I felt insightful, as when I picked out the shuffle of a quail from a welter of scuffs. Most of all I appreciated the rarity of this experience: to hunch, for long and silent minutes, over a scratch or dropping. I felt that I was back in an English classroom, slowly digesting and interpreting a complex text. "This is the first alphabet that we as a human species had to be able to read," Sarah Spaeth, an evaluation assistant, told me during a quiet moment.

Spaeth, who was visiting from western Washington, is at the vanguard of the tracking revolution. The director of conservation and strategic partnerships for the Jefferson Land Trust, Spaeth had begun tracking more than a decade earlier, after she'd attended a workshop and gotten hooked. (During one early evaluation, she'd correctly identified the pockmark produced by a male elk's urine stream.) She found tracking to be invaluable in land conservation, since it helped her confirm that wildlife was indeed inhabiting and moving between the parcels that her organization protected. "It's such an incredible tool," she said.

She is hardly the only researcher to rediscover tracking's worth. For the past decade, McFarland told me, Tracker Certification's goal had been to train and certify as many trackers as possible. Its recent growth has been exponential: In 2021, the group conducted forty-six evaluations in North America. In 2023, it ran ninety-two. Now the organization was preparing for its next phase, deploying trackers in the realms of wildlife research and conservation, and exposing academics and government agencies to tracking's validity. McFarland himself was leading that endeavor: he would soon spend ten days trailing bears with biologists from the California Department of Fish and Wildlife, in hopes of quantifying how many fawns the ursids were devouring. "There are these cool areas where skilled trackers in combination with modern technology could produce some awesome results," he said.

For tracking to truly influence wildlife biology, however, it will have to overcome reputational challenges. Tracking, after all, is a methodology reliant upon human interpretation—and humans are fallible. One 2009 study in the *Journal of Wildlife Management* found that Texas biologists surveying otters frequently mistook raccoon, opossum, and even house cat tracks for otter prints. As the authors cautioned, "issues with observer reliability . . . are potentially widespread."

Granted, technology itself isn't incontrovertible: in a study published in the *Journal of Mammalogy* in 2018, Elbroch found that tracking produced more reliable estimates of mountain lion kill rates than computer models based on satellite data. Yet concerns about tracking's accuracy still dog the scientists who employ it. "On nearly every paper that I've submitted where I've used tracking, I've had the reviewer criticize the use of tracking and be like, 'That's not really good enough,'" Sage Raymond, a Canadian wildlife biologist, told me.

Raymond is among a burgeoning class of scientists trying to take tracking mainstream. In 2021, she moved to Edmonton, Alberta, to study urban-adapted coyotes, and realized the snowbound woods were rife with "little coyote highways" of pawprints and scat. Over three years, she followed more than three hundred miles of coyote trails—up ridges and down ravines, through thickets and under bushes. ("I shredded some coats," she recalled.) The canids' tracks often led Raymond to their dens, which, she said, are "embedded in the anthropogenic matrix": in other words, near humans and our infrastructure. "People have no idea that there's a coyote with a full-on den and eight pups just twenty meters from their property line," Raymond said.

Although coyotes tend to be shy, polite neighbors, they occasionally approach humans and pets and can attack when their pups are threatened, especially when they are habituated to human food—conflicts that Raymond's research could help alleviate. Edmonton's coyotes, Raymond has found, excavate their dens on hillslopes covered in thick brush. If a likely den site occurs near, say, a school, "maybe just thin some of that vegetation, and that's probably enough to disincentivize coyotes from having a den." Tracking wild creatures helps us coexist with them.

Tanya Diamond, Ahíga Snyder's partner at the California-based research group Pathways for Wildlife, is likewise using tracking to smooth human-animal relations. Diamond and Snyder specialize in the field of habitat connectivity—figuring out how animals navigate landscapes so that environmental groups and agencies can protect those corridors. For Pathways for Wildlife, the first step is searching for deer, coyote, badger, and mountain lion sign, to determine faunal travel routes; they then install motion-activated cameras along those thoroughfares to observe animal behavior. In 2022, Diamond and Snyder found the tracks of migrating mule deer intersecting with Highway 395 near Lake Tahoe, set up a camera, and captured heartbreaking video of a deer getting pulverized by a car—a tragedy, yes, but one that helped convince the state to fund a wildlife overpass to usher the herd across the asphalt. Thanks to tracking, Diamond said, "that deer's life is not in vain."

Ultimately, tracking doesn't clash with camera traps and other technology—it harmonizes with them. Few researchers have proved that point more clearly than Zoë Jewell, a Duke-affiliated wildlife biologist and co-founder of the group WildTrack. Jewell's passion for tracking originated in the 1990s, while studying black rhinos in Zimbabwe. Her research entailed working with rhinos that had been sedated and fitted with radio collars, a process so stressful that some females slowed their reproduction or miscarried; meanwhile, local trackers mocked her fancy receivers. "All you need to do is look at the ground," they told her. She learned to track and, over the course of a decade, developed a protocol for photographing the footprints of rhinos and other wildlife and differentiating the prints by age and sex. In a 2001 study in the *Journal of Zoology*, she and colleagues showed that, by measuring and comparing the precise dimensions of footprints, they could identify individual rhinos with up to 95 percent accuracy. "If you work with trackers and interpret their knowledge, there's a huge amount of benefit to be had," Jewell told me.

Over time, Jewell applied her footprint measurement technique to other species, from cheetahs to otters to mice. She also began to work with students at the University of California, Berkeley, to develop a machine-learning program that could identify footprint photos. Although Jewell initially doubted that artificial intelligence was up to the task, WildTrack's AI program can today identify twenty species with near-perfect accuracy, and individual animals of nine of those species around 87 percent of the time. And the program is only growing more powerful as volunteer scientists, from hikers in Colorado to the San in the Kalahari, submit images of footprints to WildTrack's database using a public app. Today AI-based tracking is being used from

South Africa, where biologists are monitoring mouse populations, to Nepal, where researchers are helping herders figure out precisely which tigers are killing their livestock.

Jewell's enthusiasm for prints has not always won her favor. After she published her first rhino paper, some biologists chafed at its implicit critique of the traditional "dart-and-collar" approach. Others, she recalled, deemed tracking an antiquated form of "witchcraft." In fact, she said, the reverse is true: The sophistication and flexibility of Zimbabwean trackers make Western scientific techniques look primitive. Take motion-activated camera traps, which, for all their virtues, only collect data where you put them. By contrast, Jewell pointed out, "footprints cover the landscape; you can pick them up anywhere. Once you've learned to look down, the earth is almost like a canvas buzzing with information."

And that canvas doesn't merely offer a portrait of other species' lives. It demonstrates all we share with them. Throughout my evaluation in California, Reynerson and McFarland took pains to point out how conjoined evolutionary history manifested in tracks. The slender fingers of a raccoon recalled our own dexterous hands; mule deer prints occasioned a soliloquy on the anatomy of the hoof, whose keratinous sheaths are effectively the modified nails of our middle two fingers. "When we start to think of these animals as our cousins," Reynerson said, "we suddenly understand tracks in a different way."

If tracking demonstrates our commonalities with wild animals, it also illustrates how thoroughly we're annihilating them. At one point during the two-day evaluation, Reynerson and McFarland asked us to identify a male long-tailed weasel, his sharp face furred in a handsome black mask, that had been killed by a car. The weasel seemed to symbolize the horrors that humans wreak upon nature—and to suggest the tragedy inherent to modern tracking. To track in the Anthropocene is to document loss; as biodiversity collapses, its absence is reflected in the ground itself.

Yet tracking also indicates how much life remains. The evaluation's final day was held at a park blanketed in oak savanna, a rich biome that teemed with sign. We identified the gooey feces of a turkey, the silken trap of a funnel-web spider, the stride of a roadrunner—a "cool little dinosaur of a bird that moves around in these arid lands," Reynerson said. I pictured the roadrunner sprinting after a lizard, elongated feet stamping backward Ks upon the sand, and felt glad.

Perhaps this explains some of the growing fascination with tracking—the compulsion to reconnect to the wildlife we're losing. As tracking's ranks have swelled, its demographics have evolved. Among the evaluation assistants was Todd Cooley, a Black tracker from the Bay Area who works for an equity-minded credit union. Many minority trackers confront an obstacle course of barriers, Cooley pointed out, including the cost of attending an evaluation (mine ran $360) and the safety and transportation challenges that come with getting out into nature. Hence Tracker Certification's access committee, which has raised $25,000 to reduce barriers for participants, particularly those from historically marginalized groups. The group has held workshops at parks in Atlanta and Baltimore for Black and Indigenous trackers and other trackers of color, including biologists, open-space advocates, and educators. Even in these citified spaces, Cooley said, the biodiversity "blew my mind." Foxes deposited their fur-filled scats; mink left dainty prints in mud; beavers hewed streamside trees. Tracking brought urbanites into contact with hidden nature, and suggested new avenues for outdoor education and the preservation of green space. "I see it as this powerful piece to connect Black and brown people with each other and the natural world, and get healthier in their bodies and spirits," Cooley said.

Later, I spoke with Vanessa Castle, a fisheries and wildlife technician with the Lower Elwha Klallam Tribe, who, after beginning to track in 2021, realized it was a means of reclaiming her people's traditional knowledge—"a reconnection with the way that my ancestors used to see the world." She's since mentored dozens of young tribal members, several of whom have passed their own evaluations, a potential pathway to jobs in fields such as natural resource management. "I have an obligation to the future generations of my tribe to continue teaching them those skills that I'm learning along the way," Castle said.

In the end, I narrowly failed my own evaluation (though at

least I'd been right about the bobcat). For days afterward, I was plagued by what Reynerson called the "haunting miss": the dog prints I'd called coyote, the indistinct raccoon tracks I'd confused for fox. Yet my haplessness was beside the point. I now knew that a mourning dove's scat resembled a cheese Danish, that male tarantulas carried their sperm packets in their leg-like pedipalps, that rabbits ate and redigested their own pellets. Who could put a price on such knowledge?

At the test's end, our group gathered under an oak tree to debrief. McFarland reminded us of something that had been easy to forget, stooped in the dirt as we'd been: every track, every scat, every chew mark was the "physical extension of an animal," a flesh-and-blood creature. The mole tunnel had contained an actual mole, gorging on millipedes beneath our feet; the striped skunk tracks had been left by an actual skunk, loping down a dirt road to fulfill its secret ends. So many beings scurrying over the land—feeding, mating, killing, living—enduring everything we humans throw at them, thriving in spite of us, leaving their mark upon the world.

ERIKA HAYASAKI

Maui on Fire: When the Wildfires Came, a Young Couple Turned Toward Each Other to Survive Hawaii's Deadliest Natural Disaster

FROM *New York*

SWEATING THROUGH THE sheets, Lanz Aguinaldo rolled over in bed to reach for Isabella Lynch. It was just after 9 a.m. on August 8, and the pair normally loved sleeping in, cuddling as the sun streamed through the windows and warmed their bed. Today, the room felt like it was roasting. Isabella, too hot and uncomfortable to sleep, had been awake for thirty minutes already. A plug-in window air conditioner, which usually kept them cool during the tropical Maui summers, could not power on. Electricity in Lahaina had been out for over two hours. Outside, the wind screamed.

A mile northeast, that wind had already toppled power lines, igniting a brush fire in a field swarming with overgrown grass and weeds across from Lahaina Intermediate School, along Lahainaluna Road. Local residents had reported the fire at 6:37

that morning, but without power, television, or internet service, Lanz and Isabella had not heard anything about it.

To the northwest, Lanz and Isabella's street turned into a dead end. The only direct way out of their neighborhood, and away from the fire-ignition site, was southwest via Lahainaluna Road, two-tenths of a mile away. But the couple were not thinking about exit routes when they woke up that morning. They did not know there had been a three-acre-wide brush fire so close.

The National Weather Service had been warning of the fire threats and strong winds from Hurricane Dora, 500 miles from Maui. Forecasters predicted gusts of up to 60 mph, strong enough to move a person. The eighteen-year-olds had lived through similar red-flag warnings before. Lanz did not think today would be much different. But when she stepped outside, the hot, dusty air whipped around her.

Lanz and Isabella had met at Lahainaluna High School, where at the start of every academic year fallen mangoes blanket the red-dirt grounds. Teens tromp over the rotting fruit on their way to classes, turning the air stinky-sour. Lanz and Isabella shared two classes, English and health. Isabella had tried dating boys at the school before, and Lanz quietly watched those relationships take off, then crash like paper airplanes, all the while thinking her crush would go unrequited. Until one day over FaceTime Isabella confessed that she was attracted to women too. "Whoa," Lanz said. "That's new." On Valentine's Day, they went to a restaurant and Lanz bought Isabella flowers. By May, they were a couple.

"Half a year with the best girl I could ever ask for," Lanz wrote on Instagram in November 2021. "Being with you has been the best thing that has ever happened in my life. . . . I will continue to love you till the end."

Then a photo of Isabella with the words: SHOULD I TRIP AND FALL. Followed by a photo of Lanz that read: TRIP AND FALL FOR ME.

When the couple graduated in 2022, they dreamed of renting an apartment together. They did not have college ambitions, at least not yet. They relished their blissful lives on the island and wanted to find local jobs to continue living just like this. Priced out of rentals on Maui, Isabella moved into Lanz's parents' two-story house in Lahaina. Lanz's father, Norman, a maintenance worker for Maui hotels, had moved to the island from the Philippines in 2011, working his way up to supervisor and saving money to buy a home on Kuhua Street, across from a row of canopy-shaped warehouses. He fixed up the property himself and rented out the extra rooms to fifteen tenants.

For a long time, Lanz had not been able to afford a bed frame, so she stacked five mattresses on top of one another. Then she trained and got hired for a job as a sushi chef, and Isabella landed a job at a café and later a clothing boutique. They decorated their room with drawings, license plates, paper roses, candles, ceramic cats, and tapestries. Lanz had a secondhand shelf where she kept a tray of silver rings alongside a collection of crystals she shared with Isabella. She also had a vase with a handwritten letter tucked inside. It was from her late grandmother and was her most cherished possession. "Dear Lanz," it read. "Be good in school. . . . Be good to your parents. And always pray for me."

In May, the girls decided to take a break. Their futures were uncertain. Should they stay on Maui? Should they move to California? For now, they'd live together at the Aguinaldos' house, where they woke up hungry for breakfast on the eighth. They lit the gas stove, heating a pot of water to cook instant pancit, Filipino noodles with soy sauce and seasoning. Around noon, Lanz's nieces, nephews, and their mom came over and ate with them. School was canceled because of the winds, and the day felt like a holiday. Lanz's father had gone to work that morning, but her mother, Liza, a housekeeper for a Lahaina hotel, stayed home. The family ate and visited and chatted about the weird weather. Liza plucked her nephew's wife's eyebrows. Every few minutes, a fierce gust ripped through the streets, sending tree branches ricocheting.

Desperate for relief from the heat, Lanz's older brother, Alex, hooked up a generator, plugging in a fan along with their cell phones. The service was cutting in and out, as it had before during windstorms, with blasts arriving furiously at 67 mph. From the lanai, Lanz saw her neighbors scrambling to salvage slabs of their roof. The air smelled eerily like barbecue, and a smoky haze crept across the sky. Soon, the wind had carried the entire roof away like torn slips of paper. Another neighbor's outside walls started to peel off too.

"What kind of hurricane is this?" Lanz said to Isabella. There was no rain. Just hot, wild winds that knocked over her mom's coconut tree. It crashed to the ground.

In the past century, the area burned by wildfires in Hawaii shot up by 400 percent, according to the Hawaii Wildfire Management Organization. West Maui, where Lahaina is located, already had the highest risk of wildfires on the island. In 2018, when Lanz and Isabella were high school freshmen, a brush fire carried flames into the surrounding neighborhoods, ripping through 2,100 acres and destroying twenty-one buildings and twenty-seven cars. There were no human casualties, but it ultimately wreaked more than $4.3 million in damages.

In its aftermath, residents and wildfire experts called for power lines to be buried underground. Some warned there would not be enough firefighters—just two hundred to cover the seven-hundred-square-mile island—to take on another wildfire if it got out of control. They pushed to build another fire station six miles from the Lahaina town center. But local residents, such as the head of the West Maui Taxpayers Association, had to find the money themselves. Maui's fire chief identified a potential location. By August 2023, the residents had raised $400,000—$1.6 million short of their goal, which would have paid for the construction and shipment of a modular fire department from the mainland.

Lanz and Isabella knew that Maui was in a severe drought. In the late 1800s, U.S. businessmen built irrigation systems to feed their commercial sugarcane plantations, which eventually decimated natural ecosystems. Even earlier, European cattle ranchers moved to Maui, bringing with them nonnative grasses to feed their livestock. Lahaina's sugarcane industry began to die out in the late twentieth century. Tens of thousands of acres of former farmland sat dormant, overgrowing with these highly flammable nonnative plants. The neglected infrastructure led to Lahaina becoming even drier and hotter as climate change brought sweltering air.

"The threat of fire breaking out in these fields will likely increase due to increasing temperatures and prolonged drought periods associated with climate change," noted a 2021 wildfire-prevention report from Maui's Cost of Government Commission. It suggested fire authorities work with landowners to clear the brush. That report also warned that Maui's aboveground power lines were dangerous. If the lines toppled, experts warned, they could easily ignite the brush, especially in high winds.

By 3 p.m., local authorities had not yet made any televised announcements advising Lanz and Isabella to evacuate their neighborhood, not that they would have been able to see them without power. Maui Emergency Management Agency had not sent cell-phone alerts or posted on X warning residents to evacuate either; the department had only been noting road closures and reopenings. Isabella tried texting her mother, who lived in San Diego, but her messages failed to go through.

On Kuhua Street around 3:30 p.m., Alex went outside to check on the smoke, which was now so dense he could barely see the sun. When he came back inside, his body was cloaked in ash. Lanz's mom looked at her son and made a quick decision. Still no alerts or sirens, but the ash, she realized, must mean a fire was nearby. Liza did not want to wait for warnings that might not come. She frantically began to pack, telling her daughter to do the same. She stuffed clothes, along with two bathroom towels, into a bag. Power lines in front of their home and just beyond their balcony swung like jump ropes.

At first, Lanz didn't fully grasp the danger. She began to laugh. "Why are we packing?"

Isabella knew this was Lanz's coping mechanism under stress. She was always the one to laugh or crack jokes to lighten the mood. "We are going to be fine," Lanz reassured her. The firefighters would contain the flames. They would come back home to all their stuff. Even now, there were no warning sirens. But Lanz's mom was adamant. Her gut told her they had to go now.

Lanz had evacuated during the 2018 fire and came back home to a house perfectly intact. The firefighters put that one out, and they would put this one out too. Still, when Lanz went into her bedroom, she hugged her rack of clothes: vintage jackets, soft hoodies, a new reversible Vans puffer coat she had not yet worn. There was no time or space to take them. "I'll see you guys later," she said. Lanz grabbed the letter from her grandmother. Isabella had never evacuated because of fire before. Her heart pounded as she packed their laptops and her Nintendo Switch.

They went downstairs to the garage, loading up the family's Toyota RAV4. The heat outside made them feverish. Ash in the air stuck to every part of them, covering their clothes, flying onto their tongues, and stinging their eyes. Lanz's mom gave them black disposable face masks to cover their mouths. Lanz wore glasses and a red trucker hat with a rooster emblem. Isabella protected her eyes with the blue-light-filtering glasses she used to look at screens.

Alex got into the driver's seat, their mom in the front beside him. Lanz had her own 2005 Honda Civic, which she'd parked on the street. But she was low on gas, and they wanted to stay together, so she left the car behind. Lanz and her brother had never been very close, mostly because he was six years older. But they loved each other, and on this afternoon she began to see him as a father figure with Norman off at work. In fact, Alex would be a father soon. His wife, who still lived in the Philippines, was expecting to give birth to their first child in a month.

Lanz and Isabella sat in the back, eager to leave. But the group could not pull out of the garage yet. Other tenants in the house were still loading up a car in the driveway, blocking them in. Another five anguished minutes passed before they made it out to Kuhua Street. Lanz could see shredded garage doors from the warehouses across the street soaring through the sky.

Alex inched down the block toward Lahainaluna Road. They'd only moved a few feet when a heavy mango tree ensnared by electrical wires came crashing to the ground in front of the RAV4. If they had left a minute earlier, their car would have been crushed.

For as long as Lanz could remember, the mango tree had flourished in the yard of an elderly man with a hunched back. Isabella thought about the man, whom she always passed on her way home from work as he happily tended to his tree and garden. Where was he now? Was he trying to evacuate too? But there was no time to stop and check on neighbors.

The fallen tree did not cut off all of Kuhua Street, and Alex managed to steer around it. Every other way out would have led them down dead ends or further uphill toward the fires. Now, thicker and darker smoke billowed between homes; the fire was no longer burning the clean fuel of dry grasses in empty fields but the dirty fuel of domestic living: nail files, colanders, stuffed animals. "Holy fuck," Lanz said. "It's really close."

At 3:50 p.m., the county issued another alert. But this one also said nothing about Lahaina. Instead, it warned again about the wildfire on the other side of the island, noting the fire department was calling for "immediate evacuation of residents" in a subdivision within Kula, thirty-five miles away. The Kula fire upcountry had been raging all morning and afternoon, pulling in Maui firefighters and resources. The mayor of Maui County, Richard Bissen, had spent most of his morning in the Emergency Operations Center in Wailuku, twenty-two miles from Lahaina, his staff also largely concerned with the Kula fire, since the Lahaina fire chief reported that the Lahaina brush fire had been "100 percent contained" by 9 a.m.

Alex maneuvered past another towering tree, this one with streaks of orange flames flickering from its base, branches, and trunk. By the time they made it to the end of Kuhua Street, toward the intersection of Lahainaluna Road, cars were jammed together in a funnel trying to squeeze onto the same road out. They crawled down Lahainaluna. Visibility had worsened to the point that it was hard to see the road even with the high beams on. The windows were as hot as a stovetop. Even the recycled air in the car was hot and hard to breathe.

Moments later, yet another tree several feet away from them began to crumble and break apart, pieces of branches dropping like daggers toward their car. They sounded like shrapnel as they hit the windows. Lanz's mom screamed.

"Duck!" Lanz shouted. "Duck!"

Across Lahaina, calls poured into 911. "I live on Lahainaluna Road, right next to the fire," a young woman reported at 3:30 p.m. "My uncle is still trapped in the house. He's handicapped." Followed by a call from her mother: "He's an eighty-eight-year-old man. He cannot transport. He would literally have to be carried out. . . . The bougainvillea bush is on fire right now. I just had to leave him because I have the rest of my family in the car."

At 3:31 p.m.: "There's a fireball right behind our house." More calls. The fire was in their backyard. And in a building nearby. And up the street.

Over and over, as if waiting for permission, callers to 911 repeated the same question: "Do we need to evacuate?"

"If you feel unsafe, I would say 'yes,'" an emergency operator replied at 3:35 p.m. "But if you feel secure in your house, I would just stay home. But it's up to you."

"If we have to evacuate, where the fuck are we supposed to go?" a woman asked at 3:43 p.m. The dispatch operator told her to head to Lahainaluna Civic Center.

Herman Andaya, head of the county's Emergency Management Agency, was the person with power to decide whether or not to sound a siren on Maui that day. But he was off island on August 8, attending a FEMA disaster-preparedness seminar on Oahu. The conference took place at the Alohilani resort in Waikiki, a thirty-minute plane ride away from Maui. It was Andaya who, from Oahu, made the decision not to sound any warning sirens in Lahaina. Andaya later explained that he had thought people would have assumed it was a tsunami and headed toward the mountains—into the fire. But everyone could see the flames were already consuming the hills, roads, and now homes. A day after making that statement, Andaya resigned.

The mayor and his team at the emergency-operations center on the other side of the island also did not grasp the magnitude of the terror unfolding in Lahaina. The county issued this notification on X at 4:26 p.m.: "Residents of Lahaina Kelawea Mauka Subdivision is calling for immediate evacuation. . . . Grab your 'Go Kits' and evacuate your family and pets now." A similar cell phone alert had been sent to residents ten minutes earlier. But that subdivision was nearly a mile from Lanz's home and even farther from the traffic jam in which they now found themselves ensnarled amid scorched homes and trees.

The RAV4's windows withstood the assault of the falling branches, so they continued on. Alex was trying to head past the civic center, thirty minutes north of Lahaina to Kapalua, where their father was still at work. Norman had been dialing their cell phones to check on them, but service kept cutting out. If they could just make it to him, Norman told them over broken calls, they would be fine. Lanz thought he was being hardheaded: He didn't understand that getting out of Lahaina would be impossible.

By 4:30 p.m., Alex had steered them off Lahainaluna Road and into an outlet mall parking lot, desperately scanning for a way out. Now the stores were catching fire.

Isabella looked at Lanz. "Are we going to die?" Isabella asked her.

But Lanz told her that was not going to happen: "We're going to be okay."

Part of her just felt the need to comfort Isabella. She wanted to give all of them hope that this was not the end, and if any doubt crept into her own mind, she swatted it away. Besides, they were not far from the water. If it came to that, she told herself, they could go into the ocean.

Alex left the parking lot right before it went up in flames and headed toward the highway leading to their father's hotel. But every route was blocked by cars, debris, power lines, and fire. He turned the car around again, heading back toward the very zone they had just left. They couldn't see a clear road out that way either. They spotted the area they had parked at minutes earlier. It was in flames. "We literally came from there," Lanz said. "And now there's a fire starting."

"I don't think we're going to make it anywhere," Isabella said.

It was 4:37 p.m.

Alex maneuvered away.

"Why did we turn around?" Isabella asked through tears. "We should go back."

But there was nowhere to go. Their surroundings made it horrifyingly clear: staying put meant burning to death.

At 4:47 p.m., a man called 911. "Traffic is completely stuck," he shouted. The intersections were closed. "You need to open it up. There is no reason why. People can't evacuate this area. You need to do that. Police need to do that. People are going to die!"

More calls from trapped people came in to dispatch. At 4:49 p.m. a woman cried: "They closed the roads." She and her companions were heading back toward the fires. "Where are we going then?"

"Where are you?" the operator asked. "Everybody should be pushing you toward Lahaina Civic Center."

The woman said police had directed them down a dirt road. But two men had closed and locked the gates to it because of downed power lines and blocked the road with a truck. She told the operator the men did not appear to be police or authorities. Now, she said, the police who had first directed them were gone. "Help us."

Meanwhile, at 5:03 p.m., another county alert went unreceived by Isabella, Lanz, and her family. "Lahaina fire flare-up forces Lahaina Bypass road closure," it read, seemingly still oblivious to the reality on the ground. "Shelter in place encouraged."

Alex drove through a hailstorm of embers that sounded like they might crack the window glass. He tried to make it to his father's hotel. But the highway was also blocked. These blockades remain under investigation, and police have released body-camera footage showing officers helping residents evacuate. But Lanz and Isabella specifically remember police blocking the way. Lanz took a video of a police car, blue lights glowing.

So many lines had fallen. Everyone on the ground, including law enforcement, seemed to assume the wires were live and dangerous, as power-company press releases had advised. Some residents moved the lines with branches. But Hawaiian Electric later said it had de-energized power lines in the area by 7 a.m. after learning about the morning fire.

Like so many others, police redirected Alex to Front Street, along the ocean. Maybe, they reasoned, if they made it to Liza's workplace—a hotel on Front Street—they could shelter there. A week earlier, Isabella and Lanz had cruised the block, stopping at the McDonald's after work for fries and Diet Cokes. They had recently dressed up in pink and watched the *Barbie* movie down the block. Now, Front Street too had become an inferno. The McDonald's, and the Bank of Hawaii, and the theater—all of it burned. People trapped in cars, on foot, and on bikes all desperately searched for somewhere to go as the flames devoured everything in sight. They could not make it to the hotel, and even if they did, the hotel would not be safe.

Alex put his head down on the steering wheel and started sobbing. Lanz could not believe it. Her brother never cried. She had never even seen him scared. Lanz did not want to cry. There was nowhere else to go, she told herself, but the ocean.

"Let's just get out," Lanz said. "Let's go into the water. We can do this. Let's just get out right now." Lanz started gathering whatever they could take with them into the sea.

Isabella, able to get a brief signal, texted her mother. She told her they were trapped in a fire. But there was nothing about it on the national news yet. Isabella's mom checked her daughter's location on Find My Friends and told her daughter to please get into the water if she had to. She knew Isabella was an expert swimmer who had grown up practicing in the ocean.

"I'm scared," Isabella texted back. "I don't want to die." Isabella told her mother she loved her. Then cell service went out. Her mother checked her location tracker again. The last ping was on Front Street.

The wind whipped embers at them as they left the RAV4, leaving their laptops, documents, and jewelry behind. They climbed over a stone ledge, descending down a bank of jagged rocks. Lanz helped an older man in a white tank top as he clambered down the rocks. He didn't seem strong enough to swim and stayed behind where fire would soon lick the edges of the shore. They went into the water as the waves boomed like thunder.

The sea was unlike any water Isabella had ever tried to swim in. It crashed over their heads. All four grabbed onto a palm frond to stay together. It took all of their strength to hold on amid the fury of wind and waves.

The foursome shared the two towels that Lanz's mother had brought, wetting them with seawater and draping them over their heads. Each gust brought a shower of fiery embers and air scorched with hot ash. The towels blocked their faces and shielded their eyes from the smoke. Every filtered breath took effort. They could not see one another, but they could hear one another.

"Are you holding on?" Lanz's mom asked every few minutes.

"Yeah," Lanz replied.

"Are you still there?"

Lanz's mom clung onto her brother. Lanz knew her mom, fifty-one, was not a strong swimmer. They tried to stay close to the shore, their feet tiptoeing and bouncing along the ocean floor. The palm frond whipped around in the waves like a battle rope, and they were yanked around with it.

How long are we going to be here for? Lanz asked herself. Surely, someone would rescue them soon. There were so many people around them, floating, swimming, trying to shelter on the rocks. The Coast Guard. Emergency crews. The military. Someone was definitely coming to help. It was just after 6 p.m.

At 6:03 p.m., the mayor appeared live on television. "I'm happy to report that the road is open to and from Lahaina," Bissen announced in an interview with KITV news. "We had some challenges early today trying to get people in and out of Lahaina. That road opened at about five p.m."

Bissen did not indicate he had any idea people were trapped on burning streets or had fled into the water: He said only that he didn't have an update on how many people were using the evacuation center at Lahaina Civic Center and didn't know how many structures had burned. "Our update is actually going to come in at about seven p.m.," he said, "and we'll get all the reports at that time."

"What should residents there know about tonight's fire?" the newscaster asked.

The mayor encouraged people to watch the live news and visit Maui County's website and social media. "We've been putting out a pretty steady stream of updates throughout the day." And, he reiterated, "Don't touch any downed power lines. Don't try to clear the road yourself."

Mayor Bissen would say at a press conference three weeks later that he did not know who was in charge in Lahaina that day. Reports of 911 calls and police dispatches do not appear to have been relayed to him or his team at emergency operations in those unfolding hours, and no one from his office has been able to explain why. Communication seems to have collapsed in the fast-moving chaos. At the emergency-operations center, Bissen had been waiting for that early-evening update from the Lahaina fire chief. It did not come. That was when he began to realize something must be wrong. He did not even know anyone had died until the next morning.

At the off-island FEMA event, director of the Hawaii Emergency Management Agency Major General Kenneth S. Hara went on live news on the evening of August 8. He said his office was providing traffic control and security support to Lahaina from the National Guard. But he had no idea the town was already decimated or that people were stranded in the water. "I thought everyone had gotten out safely," Hara would later tell reporters. "It wasn't until probably the next day that I started hearing about some fatalities."

A short plane ride away from the FEMA event, the girls bobbed in the water beneath their towels. Three hours passed. Still no rescuers. Isabella knew there was a vast military force in Hawaii. Why had no one showed up yet? *Had the fire wiped out the entire island?* she worried. *Had all of the firefighters perished too? Where were they?*

From her vantage point, Isabella could peek from beneath the towel and see Lahaina's historic banyan tree on fire. The harbor was engulfed. Boats burned. A giant piece of sheet metal from a nearby restaurant's roof hurtled into the water, scratching Isabella. Every car that caught fire or exploded brought more black, suffocating smoke.

Other people in the water had started crying out in pain. Sea urchins were impaling their feet. Isabella and Lanz knew sea-urchin stings can be venomous. Isabella wore black sandals. Lanz wore white Crocs. Both young women loved collecting Jibbitz charms for their Crocs: anime characters, butterflies, wings. In the waves, one of Isabella's sandals kept floating away, but they immediately grabbed it, trying to keep her feet covered.

From beneath the towels, they could hear people screaming, crying, coughing, wheezing, running out of breath. Others around them did not have towels or masks to block the smoky air. Lanz worried most about her mom. She had to stay strong for her.

Another hour passed. Some people around them decided to try to swim to Baby Beach, a shoreline about a quarter mile away. It was a dangerous journey in this surge. So many were exhausted from the water and smoke. Isabella thought she could make it and that maybe Lanz could too. But Isabella knew Lanz wouldn't want to leave her mom. And Isabella didn't want that either.

"Just breathe," Lanz told Isabella. "We can do this."

She told her mom, "We have to hold on. We'll fight through it. Be strong."

"I love you," Lanz told Isabella. "I don't want to lose you. I don't want to lose my mom."

"I love you too," Isabella told Lanz. She realized that if they made it out alive, she was going to end the break they were on. They had to be together.

Now, five hours had gone by. Isabella thought about her sister. She worried about her mother in California, who she knew was fretting about her and staring at her phone, waiting for a reply. She thought about the last time she had seen them, how happy she felt. Would she ever see them again? She wondered if she should have gone to college.

Isabella felt herself drifting off to sleep. She was so exhausted. Then she heard a voice in her head: *You're going to be okay.* She hoped it was right. But the voice washed over her again. It helped her gather the strength to fight off sleep. "Breathe," she told herself. "Breathe."

Every now and then, the wind would let up a little. The smoke seemed to clear a bit. In those moments, Lanz and Isabella inhaled deeply, relieved, as if breathing fresh air. Alex pulled a plastic bag out of the backpack. *A Filipino butter muffin!* They each had a bite.

The water turned frigid as darkness fell. The salty sea stung the cuts covering Isabella's legs and feet. Lanz shivered. Beneath the towels, she listened closely for the sounds of her loved ones breathing. And she cried. Death was all around her. She could hear it. Isabella assumed people were dead already in the water, on the rocks, their bodies hidden by smoke. But she would not die today. They would not.

Over the past couple of years, Lanz had captured so many moments on her iPhone, now stuffed inside of a backpack floating in the waves with them. Some of these photos she would share; most she kept for herself. Lanz had a clip of herself from months ago, holding her mom in her arms and spinning her around as her dad lounged behind them. I LOVE YOU NANANG, she wrote under it. She had even filmed their escape from the fires, right up until the moment they got into the water.

Earlier in the year, Lanz posted a photo on TikTok of a Maui street in the dark under a rising orange sun. She wrote, PHOTOS IVE TAKEN THAT MAKE ME APPRECIATE LIFE. Now, six hours in the water had passed. Soon the sun would rise. Lanz inhaled.

Just after midnight, the fire seemed to have burned itself down to the shoreline. It was beginning to subside. People who had managed to stay alive in the water began climbing back onto the rocks. Lanz pulled herself up the bank. She spotted the older man in the tank top whom she had helped earlier that day. He looked like he was lying down. "Hey, sir," she said. She tried to wake him, then realized he was dead.

One survivor took videos of what was left. Carcasses of hundreds of cars. A small, burnt body beneath a charred vehicle. A dead dog, singed to ash. Remains of people still in cars. A lifeless man, face up on the shore, his shirt lifted up above his belly. A dead woman splayed across the rocks, her hair covering her face. Bodies floating in the waves. One wore red.

The wind had calmed a little. Still, they kept the towels over their faces. They walked down the blackened road until they got to their burned-out RAV4. A yellow suitcase in the back seat had partially melted but otherwise remained intact. Their phones, surprisingly, still worked. Isabella's Nintendo Switch didn't make it. But the letter from Lanz's grandma did.

At around 1 a.m., still huddled on the rocks, Lanz and Isabella began to see lights in the water. They thought it was the Coast Guard. Everyone began waving and flashing their cell phone lights to get their attention. The boats sailed past them three times. Isabella gave up hope that they were going to come for them.

Thirty minutes later, the boats sent swimmers on surfboards to survey the shore. Isabella remembered nearly 150 people going into the water. Now, she figured there were about 70 left. Some had made it to Baby Beach. Some must have drowned trying. Others died near the shore. Still others must have managed to escape to somewhere else.

Isabella wondered, *Did they plan to take everyone who was left, one by one, aboard a boat via a surfboard?* They had just gotten out of the water, and it was so cold and dark. They did not want to get back in. Isabella's phone had 2 percent of its battery life left. She tried to call her mom several times before finally getting through. She heard her sobbing with relief. Isabella wept too.

Finally, a fire truck arrived. Eleven hours had passed since Isabella, Lanz, her mom, and her brother had fled their home. It was the first fire truck they had seen. "We need able-bodied people who can walk to our trucks," one fireman shouted.

Isabella couldn't believe their lack of urgency, their terse tone. She spotted an older woman who couldn't walk being loaded into a grocery cart. Another man was hyperventilating. He had been helping people climb the rock wall, and his lungs were giving out on him. The firefighters did not seem concerned. "It was not like they were rescuing people who had just been through a near-death experience," Isabella would later say. "They were just like we were a little job, like chores that they had to do."

They were grateful, of course, but also confused. Maybe the firefighters had gone through too much already? Isabella, Lanz, and her mom and brother climbed into the open back of the truck. As it drove away, they realized that the town was totally leveled. Buildings were still shooting flames as they sped past. Firefighters yelled at them to duck when they encountered power lines. Lanz and Isabella looked around in shock. Lahaina was a graveyard.

Lanz began to worry about her uncle, nieces, and nephews. Had they made it out okay? And their neighbors? The ones whose roof blew off? Lanz adored their seven-year-old grandson, Tony. She had known him since he was in diapers. He would come over and ask her innocent questions. Lanz would answer, sharing food and jokes with him. She later learned her father's tenants had gotten out safely, and so too her nieces and nephews. But her neighbors did not make it. Grandfather, grandmother, mother, and Tony all died trying to escape.

They headed toward the Lahaina Civic Center. But as soon as they arrived, officers stopped them. The fire was too close to the shelter. They had to evacuate to another one, at Maui Preparatory Academy, with seven hundred others. When they arrived, there were no blankets; they wore their sea-soaked clothes. Isabella shivered and Lanz woke up crying. They did not shower. Still covered in ash, Isabella's goldish-brown hair had turned black. Even their eyebrows were stained.

The next morning, Lanz's mom got a cell signal and made contact with Lanz's father. He had spent the night terrified that their call before getting into the ocean was the last time he would ever hear their voices. He drove them to one of the hotels where he worked.

Days later, they would return to their home on Kuhua Street. Everything—Lanz's car, bed frame, bedroom, clothes, the sofas, the lanai, the entire home—had been reduced to a pile of rubble.

Two days after the fire, Isabella asked Lanz to be her girlfriend again. Lanz said yes. They flew to San Diego to stay with Isabella's mother. Meanwhile, Lanz's family stayed at the Sands of Kahana hotel on Kahana Beach, nine miles from their burned-down property. Everything was chaotic. Lanz and Isabella did not want to burden Lanz's parents. Lanz's mother now had no job. The Lahaina hotel she cleaned had shuttered. Lanz's father had lost all of his tools in the fire. Now, they slept in a hotel room with sheer orange drapes, orange cushions, and an orange sofa. It had a kitchen stocked with noodles, peanut butter, Hawaiian rolls, and a rice cooker.

In 2022, Hawaii had the highest cost of living of any state, including California and New York, and renting on Maui has only gotten harder for its middle-class residents as off-island real-estate investors have bought up homes and turned them into vacation rentals. Lanz's family were among 6,800 fire survivors now living in thirty-six hotels on Maui. The tourists would return soon; the governor declared West Maui open again exactly two months after the fire. Survivors still did not know where they would go next.

Lanz's father said his home's last appraisal before the fire came to $1.2 million. But he still owed $715,000 on his home loan. His insurance would only cover $500,000. This meant he would have to take out another loan to rebuild it, not to mention the labor costs. But he was determined to get his place back, even if it meant building it himself. Why not leave? His voice broke when he answered: "Because I love Maui." On the carpeted floor under the kitchen counter, he had already begun collecting tools to rebuild, bringing a new one back each day: a finishing-nail gun, a power drill, a framing gun, a circular saw, a construction ruler.

Meanwhile, Lanz and Isabella remained in a state of disbelief. "Before, we were just grateful to have been saved, and we still are," Isabella said. "But now we're just like, 'Why wasn't it sooner? How could we have gone through this? Why? Why were we all trapped? Why were we not evacuated?' Just so many different questions. And a lot of sadness has turned to anger now. Like, why didn't they do something to avoid this happening? I mean, our whole town burned down. They could have at least like tried harder to get the people out. In that kind of situation, even if there's downed power lines, you let people through because so many more people would have survived."

The same questions now sit at the center of investigations by Congress, the Hawaii attorney general's office, Hawaiian Electric, and Maui County, as well as over two dozen lawsuits, all probing what happened on and before August 8 to cause the deadliest U.S. wildfire since 1918, killing one hundred people ranging in age from seven to ninety-seven.

After a few weeks in California, Lanz flew back to Maui. She felt guilty for being away from her parents. They needed to figure out their insurance and income, how to solicit donations and begin to rebuild. Isabella helped create a GoFundMe page for them online. But Lanz wanted to do more.

It was hard leaving Isabella, even for a week. But they had survived the fire, the wind, the ocean. Their relationship, Lanz was confident, could survive this separation too. And by October, the two had reunited. Lanz posted an image of them kissing and wrote, HAPPY BIRTHDAY MY BEAUTIFUL GIRL. I LOVE YOU TO THE MOON AND BACK. Lanz also TikToked herself in her "pumpkin patch fit" jean shorts and an olive button-down, to which Isabella replied, "u my pick of da patch."

Recently, Isabella posted an old photo of Lahaina glowing orange at sunset. It was taken from her home on Kuhua Street. "I miss my house," she wrote in one steady stream of thoughts. "i miss my girlfriends car i miss my job and walking to work and walking all around lahaina i miss foodland i miss front street and the banyan tree and my elementary school . . . oh lahaina i just miss everything about you so much and im so sad and angry i miss my home."

By November, Lanz and Isabella had managed to get their old jobs back. The boutique and the sushi restaurant, neither of which had been damaged by the fire, reopened. The couple found a small apartment together in Honokowai, about ten miles from their burned-down home. Lanz's parents are still rotating among hotels until they can move into a rental house in February.

Lanz is posting again. Lip-syncing, dancing, posing in outfits. She recently posted a photo reel of their relationship with the caption SHE KEEPS ON LOVING ME AND I KEEP WONDERING WHY, one of her favorite love songs by the Red Clay Strays playing over the images. But off-camera, Lanz often thinks back to that day. The strength she mustered in those horrifying moments. She had never been the serious adult in the room, the person ready to take charge. But on August 8, she grew up faster than she had ever wanted to, her words to her mom and Alex and Isabella in the water a mantra moving forward: "We can do this. We have to hold on."

FERRIS JABR

The Whale Who Went AWOL

FROM *The New York Times Magazine*

ON APRIL 26, 2019, a beluga whale appeared near Tufjord, a village in northern Norway, immediately alarming fishermen in the area. Belugas in that part of the world typically inhabit the remote Arctic and are rarely spotted as far south as the Norwegian mainland. Although they occasionally travel solo, they tend to live and move in groups. This particular whale was entirely alone and unusually comfortable around humans, trailing boats and opening his mouth as though expecting to be fed. And he seemed to be tangled in rope.

When a commercial fisherman named Joar Hesten got a closer look, he realized that the whale was in fact wearing a harness: one strap girdling his neck and another gripping his torso just behind his flippers. Hesten contacted a local scientist, and word eventually reached the Norwegian Directorate of Fisheries, which dispatched an inspector, Jorgen Ree Wiig. After several failed attempts by Wiig and a colleague to free the beluga while on board a dinghy, Hesten put on an immersion suit and plunged into the water. Though the whale was not quite as hefty as an average adult male of his species, he was still a formidable presence, by best estimates close to fourteen feet long and about

2,700 pounds. Swimming beside him, Hesten managed to un-clasp one of the straps. Together, they used a grappling-hook-like device to remove the rest of the stubborn harness.

A few days later, the beluga followed a boat to Hammerfest, one of the northernmost towns in the world, where he took up residence, frequently interacting with people in the harbor. News of the friendly white whale spread quickly. In early May, a video of the beluga went viral, eventually earning a spot on *The Ellen DeGeneres Show*. In it, several young women stand on a dock in Hammerfest, speaking excitedly with their hands outstretched just above the water. The beluga levitates to the surface in an up-right position, as smooth, plump, and silent as a balloon. There is something in his mouth—something rectangular. "Oh, my God!" one woman exclaims as the whale returns a smartphone her friend dropped in the sea. The women cheer and caress the whale, whose mouth continues to hang open. Later viral videos would show him stealing (and returning) a kayaker's GoPro and playing fetch with a rugby ball. By midsummer, he had become an international celebrity, drawing large groups of tourists.

All the while, marine experts had been speculating about the whale's origin. Clearly this animal had spent time in captivity—but where? The first major clues came from the harness: One of its plastic buckles was embossed with the words "Equipment St. Petersburg." And it appeared to have a camera mount, hinting at reconnaissance of some kind. The beluga also knew how to closely follow boats and had a habit of wrapping rope around propellers, which could suggest specialized training. As several experts told me-dia outlets at the time, the whale had most likely escaped from the nearby Russian Navy. Based on a poll of more than twenty-five thou-sand respondents, the Norwegian Broadcasting Corporation chris-tened the beluga Hvaldimir, a portmanteau of *hval*, the Norwegian word for whale, and the Russian name Vladimir.

The military conscription of a beluga whale might sound like a conceit plucked from less-than-convincing spy fiction, but it is actually a well-documented practice. Since the 1960s, Russia and the United States have trained dolphins, seals, and other marine mammals to assist their naval forces by tagging enemy divers, detecting mines, and recovering items from the seafloor.

Satellite photos of Russian naval bases near Murmansk, not far from the spot where Norwegian fishermen first found Hvaldimir, reveal the type of sea pens often used to hold belugas. Audun Rikardsen, a professor of marine biology at the Arctic University of Norway, told me that international contacts have since confirmed that Hvaldimir belonged to the navy.

In the years since Hvaldimir first entered the global spotlight, the very qualities that make him so endearing—his intelligence, curiosity, and charisma—have put him in perpetual danger. While traveling along the coasts of Norway and Sweden, he has inadvertently hooked himself on fishing lines and suffered multiple gashes caused by boat strikes. Incessant chewing of ropes and chains has worn his teeth to nubs. Overzealous spectators have swarmed him for photos, prodded him with brooms, and thrown rocks in his vicinity to draw his attention. Some Norwegians have threatened to seek warrants to shoot and kill the beluga because he has damaged salmon farms or other underwater structures.

Hvaldimir is now at the center of a dispute over his welfare. Although he has become more independent since his early residence in Hammerfest, he has not completely relinquished human companionship. He has retained enough survival skills to feed himself, yet he has also ventured into warmer waters where there are no belugas, insufficient food, and numerous threats. Even as he swims freely through the ocean, he is caught in a tangle of conflicting human ambitions, some noble, others misguided, nearly all distorted by inadequate understanding. Whether to intervene, and how to do so, remain contentious subjects among scientists, activists, and government officials.

Many advocates would like to see Hvaldimir reunited with wild belugas or at least moved to a nature reserve. But rehabilitating a formerly captive whale is nothing like the triumphant leap to freedom in *Free Willy*; it's more like helping a severely traumatized victim of abduction reintegrate with society. For creatures of such size and sentience, confinement to relatively tiny, sparse, and lonely cells exacts a heavy physical and psychological toll. Like Hvaldimir, many captive cetaceans are in-between creatures: born to whales but raised by humans, not quite domesticated but no longer wild, suspended somewhere in the middle of instinct and compliance. Hvaldimir is a living bridge between their circumscribed existence and the nearly limitless one from which they were barred. What happens to him now—whether he becomes a rare example of successful rewilding, transitions to a more sedate life in a sanctuary, or meets a tragic end like so many of his predecessors—will influence efforts to liberate the thousands of cetaceans still in confinement today.

Wherever Hvaldimir goes, he is followed by a small but passionate entourage of human defenders and devotees. One individual among them has become especially prominent and controversial: Regina Crosby Haug, an American filmmaker whose entanglement with Hvaldimir is largely a product of circumstance.

In 2019, after rekindling a relationship with her high school sweetheart—a Norwegian man who came to her Idaho hometown as an exchange student—Crosby Haug started splitting her time between Southern California and Norway. When she learned of Hvaldimir, she decided to take advantage of their proximity and visit him in Hammerfest, where she hoped to collect some interesting footage. Their first meeting took place near a salmon farm. "He swam up to our boat full of people with a fish he had caught and gave it directly to me," Crosby Haug recalls. "I was blown away. I couldn't believe he could make that kind of connection. I thought to myself, I think I just made a friend."

The more Crosby Haug learned about Hvaldimir, the more she feared for his future. In addition to the daily dangers he faced in the water, there was little regulation of the crowds that flocked to see him. And some individuals in the oceanarium industry, Crosby Haug heard, might have their eyes on him. Over time, what began as a short upbeat video grew into a feature-length production—and a life-consuming mission.

In the fall of 2019, Crosby Haug created an informal advocacy group called Friends of Hvaldimir to raise awareness of the beluga's plight. The following summer, she officially founded the nonprofit OneWhale, which is dedicated to protecting Hvaldimir. Several esteemed cetacean scientists—including Ingrid Visser, Diana Reiss, and Roger Payne—joined the organization as advisers.

Other people in Norway were falling for the whale too. In July 2021, Sebastian Strand, a burly, softhearted, twenty-four-year-old diver and graduate student in marine biology, chanced upon Hvaldimir swimming circles in a harbor in Vevelstad, not far from his hometown. As he walked along the docks, Hvaldimir surfaced and approached him. Strand immediately called his friend and canceled their planned fishing trip. Instead, he spent the next eight hours interacting with the inquisitive whale, eventually entering the frigid water in just swim trunks and a shirt. By early 2022, Strand was working for OneWhale full-time, in tandem with its network of volunteers.

Strand has since devoted nearly every day to watching over Hvaldimir and assessing his health, following him by car and boat, never knowing exactly where he might have to travel next and often sleeping in a vehicle, at a hostel, or on a kind local's couch. Depending on the situation, his work has entailed public outreach, crowd control, and first aid. Over the past two years, Hvaldimir has very likely formed a stronger bond with Strand than with anyone else. "Hvaldimir has opened my eyes to a new level of animal intelligence," Strand told me. "Over the time I have spent close to him, he has gone from a curiosity with a potentially tragic background to an individual I care about deeply. In many ways, I see him as a person."

OneWhale's efforts fill a vacuum created by the ambiguity of Hvaldimir's situation. Because he is a formerly captive animal living in the wild, it's not clear who, if anyone, should be responsible for him. Russia has never claimed ownership of Hvaldimir, nor has anyone else. No prominent international animal rights or conservation group has volunteered to oversee his welfare. In May 2019, when Hvaldimir was noticeably emaciated, a research group called the Norwegian Orca Survey set up a program to feed him frozen herring by hand. By fall, fecal samples indicated that Hvaldimir was learning to catch live fish for himself. Since then, the Directorate of Fisheries has maintained a position of mild indifference, insisting that Hvaldimir is a wild whale and can fend for himself.

When Crosby Haug founded OneWhale, she already suspected that chasing a whale through the ocean and trying to keep him out of trouble would not be a sustainable strategy. In parallel, she began pursuing an alternative solution: recapturing Hvaldimir in order to save him.

The controversy surrounding Hvaldimir is part of a much larger debate concerning the ethics of cetacean captivity. Humans have been wresting whales from the ocean and keeping them in tanks since at least the 1860s, when P.T. Barnum exhibited live belugas in Boston and New York. At the time, many Westerners perceived whales as "monsters" that could be hunted, displayed, and discarded without misgivings. Since then, research has established that cetaceans are self-aware, empathic, and highly intelligent beings, many of whom form lifelong relationships and maintain genuine forms of culture. A growing number of countries, including France and Canada, are now banning all future cetacean breeding and captivity. Some aquariums and marine-mammal parks have already agreed to retire and rehabilitate their orcas, belugas, and dolphins.

Many of these changes have been spurred by increasing social pressure. In the past three decades, and especially since the harrowing 2013 documentary *Blackfish*, the public has become much more critical of cetacean captivity, which can result in both deformities and behavioral abnormalities. While there are few verified accounts of wild orcas harming humans, captive orcas have attacked trainers many dozens of times and in several cases have killed them. Yet an estimated 3,600 whales, dolphins, and porpoises still live in confinement around the globe. Since at least the mid-2000s, scientists, conservationists, and some oceanariums have been trying in earnest to establish what many experts agree are necessary and viable alternatives to standard captivity: open-water sanctuaries. Animals who can't transition to life in the wild can live out their remaining years in a protected semi-wild space that dwarfs any tank—at least in theory.

An ideal cetacean sanctuary should be sheltered but still part of the ocean; it should be large, remote, and untrafficked, yet still small and accessible enough to staff and manage. In other words, exactly the kind of place that humans like to keep for themselves. This was the predicament the Whale Sanctuary Project, an

American nonprofit, confronted when it began searching for a site to establish a haven for orcas and belugas. Following years of staunch opposition from local residents and fishermen, the organization finally found one bayside town in Nova Scotia that welcomed their proposal for a hundred-acre sanctuary. They are currently acquiring the necessary permits, a process that has spanned more than two years, and they don't yet have any whales confirmed for rehoming. In 2016, the National Aquarium in Baltimore announced plans to develop a sanctuary for its six Atlantic bottlenose dolphins, but it has also encountered numerous hurdles, including projected storm surges and other dangers that climate change will ultimately impose on captive creatures with such long lifespans. "We took a hard look at the Florida Keys," says John Racanelli, president of the National Aquarium. "But the hurricanes of 2017 opened our eyes to the fact that we'll likely be caring for a succession of dolphins across many decades. Our facility still needs to be functional in 2100."

Merlin Entertainments, a global operator of theme parks and other attractions, has been developing an eight-acre beluga sanctuary in Iceland since 2012. In 2019, Merlin and its various partners transported two belugas from a Shanghai aquarium to a bay on the remote island Heimaey—the same bay that housed Keiko, the orca that starred in *Free Willy*, during his attempted rehabilitation in the late 1990s. As with parallel efforts, the environment has been problematic, especially in winter. Jeff Foster, a cetacean-welfare expert who worked with Keiko in Iceland, recalls wind gusts up to two hundred miles per hour and strong waves that displaced nets. Equally challenging has been the complexity of cetacean psychology. One of the belugas is struggling to adjust to life in the sea, perturbed by the unfamiliarity of currents, tides, and even rain. Because of her hesitancy, combined with harsh winter conditions and health troubles, the belugas have spent most of the past four years living in an indoor pool in a landside facility.

Within weeks of meeting Hvaldimir, Crosby Haug began contacting every cetacean sanctuary she could find, but none were willing or able to house a fugitive beluga. Eventually she consulted Ric O'Barry, a renowned environmental activist. In a previous life, O'Barry captured dolphins for the Miami Seaquarium and trained them to perform in the 1960s TV series *Flipper*. In 1970, one of the show's starring dolphins died in his arms after failing to resurface for air—an incident he interpreted as suicide. (Unlike most mammals, cetaceans must consciously choose to breathe.) The experience changed him forever.

O'Barry and his son Lincoln have since established what are, in some respects, the most successful cetacean sanctuaries in the world. Working with the Indonesian government and a local nonprofit, the O'Barrys created two permanent facilities in Bali and Karimun Jawa for the retirement, rehabilitation, and release of former dolphin performers. The facilities, situated in sheltered coastal areas, consist of wooden sea pens in which dolphins unlearn their captive behaviors and develop the skills they need to survive in the wild, like hunting and deep diving. Since 1973, the O'Barrys have rehabilitated and released more than twenty dolphins in various parts of the world, a majority of which they are confident reintegrated with wild pods.

Crosby Haug initially asked Ric O'Barry to spearhead the effort to save Hvaldimir so that she could focus on her documentary, but he declined because he was too busy with his own projects and skeptical that the Norwegian government would offer the necessary permissions and assistance. Instead, he encouraged Crosby Haug to lead the campaign herself. Shortly thereafter, she approached the mayor of Hammerfest about creating a new reserve to rehabilitate and release Hvaldimir and other belugas. After all, Hvaldimir had already enchanted the citizens of Hammerfest, and the town itself was surrounded by pristine Arctic habitat. The mayor connected her with Katrine Naess, a destination developer for a local tourism company. "Regina was really good at selling this as an opportunity that could be a win-win," Naess says. "Everyone wants to be the town that saved Hvaldimir." In August 2022, Naess, Crosby Haug, and several colleagues founded a nonprofit called the Norwegian Whale Reserve, which has been trying to realize their mutual ambitions ever since.

The project's proposed location is a two-hundred-acre fjord

about twenty miles southwest of Hammerfest. Turning it into a reserve would require stretching thousands of feet of net across its mouth and securing it all to the seafloor. Such nets need to be flexible enough to accommodate waves and tides, while also remaining taut enough that they don't bunch up and trap the animals. You might think a whale or dolphin would swim or jump over any barrier level with the ocean's surface, but most cetaceans seem to have a mental block that prevents them from doing so. (The climactic leap in *Free Willy* was accomplished by strapping an animatronic whale to a rocket launcher.) "This is not a sea pen," Naess says. "Our ambition is to make a beautiful open-sea reserve where there is no civilization, no traffic—ideally just pure nature. We want to set an example for the rest of world."

Norway is a somewhat unlikely choice for a cetacean sanctuary. It is one of only three countries, along with Japan and Iceland, that continue to engage in commercial whaling. Polling suggests that most Norwegians have consumed whale meat at some point and that less than a quarter of the population supports an immediate end to whaling. There's also a major bureaucratic obstacle: Norwegian law stipulates that a wild whale cannot be held captive unless it is part of a zoo or scientific study, neither of which is particularly compatible with a model sanctuary. Although the Hammerfest municipal council has not yet officially sanctioned a reserve, it voted in favor of conducting preliminary environmental tests of the proposed site. So far, the results are encouraging, indicating exceptionally clean water.

The largest sanctuaries in development can each hold ten to twenty cetaceans at most, a tiny fraction of the world's captive population. I asked Lincoln O'Barry why it was taking so long to do so little. He explained that among all performing animals, whales and dolphins—in particular, orcas—are uniquely lucrative. The estimated market value of a single captive orca is between $1 million and $10 million, many times the typical selling price of an elephant, a tiger, or a great ape. "I don't see any aquarium giving up that kind of asset," he told me. "There have been sanctuaries and releases going on for all kinds of terrestrial species, and the whole oceanarium industry is trying to make it sound like it's not possible. Whales and dolphins are basically the last animals on Earth that have to perform seven days a week until they die while living in a completely barren box without even a rock to hide behind."

For cetacean welfare advocates, each passing year without an adequate network of cetacean sanctuaries permits the possibility of further tragedy. Last March, the Miami Seaquarium announced a legally binding agreement to relocate a female orca named Lolita, who had been in captivity since 1970, to a reserve in her native Salish Sea. Five months later, while still occupying an eighty-foot-long tank in Florida, Lolita died from kidney failure. Her ashes were packed into a white box painted with an exact replica of her tail and topped with cedar boughs, flown to Washington State, and given to members of the Lummi Nation, who consider Lolita, or Sk'aliCh'elh-tenaut as they call her, to be their relative. In a private ceremony in September at a sacred site, Sk'aliCh'elh-tenaut finally returned to the sea.

Belugas were once thought to have a maximum lifespan of about fifty years. The latest research suggests that they can live for close to a century. Hvaldimir's physical characteristics indicate that he is probably a young adult between twelve and twenty years of age. Had he remained in the wild from birth, he would have spent his life traveling the seas with his kin in groups of two to ten and herds of more than a thousand; communicating through complex vocalizations that scientists have only begun to decode; and learning how to be a whale from his elders. He would have had a family, a dialect, and the beluga equivalent of a name—a signature contact call—rather than another species' pun. Instead, he was probably abducted as a calf, severed from cetacean culture, and forced to undergo military training in exchange for food. In all likelihood, he either escaped from a damaged sea pen or was accidentally separated from the navy during a training exercise.

Last spring, perhaps because of a muddled migratory instinct or the drive to find a mate, Hvaldimir began an unprecedented journey south. In the past, following his departure from Hammerfest, he primarily lingered around remote salmon farms in northern Norway, where he learned to hunt the wild fish that gathered to eat spilled food pellets. He would often stay in a single location for months, allowing Strand to mediate relations between Hvaldimir and local salmon farmers. By April, Hvaldimir was speeding down the coast of Norway, rarely staying in any one place for more than a few days. On May 19, he reached Oslo. A few days later, he was spotted off the coast of Sweden—the first time since 2019 that he had crossed another country's borders. All the while, he was swimming farther from food-rich and relatively tranquil waters toward larger and more dangerous harbors. At times, he entered industrial zones and murky canals—exactly the kinds of places in which solitary cetaceans tend to get stuck.

Even before his unexpected voyage, Hvaldimir's behavior had been changing. Beyond his newfound wanderlust, he became less interested in following boats and hanging around humans compared with previous years. He appeared to be growing wilder all on his own. Because of this evolution, the escalating threat to Hvaldimir's well-being and the absence of a suitable sanctuary, OneWhale revised its strategy. In partnership with the Norwegian animal rights organization NOAH, OneWhale is now petitioning Norway's government to relocate Hvaldimir directly to Svalbard, an archipelago about halfway between the Norwegian mainland and the North Pole, with the nearest resident population of wild belugas. OneWhale hired Jeff Foster to write a detailed report explaining how to transport Hvaldimir by ship or plane. The Norwegian fisheries director, Frank Bakke-Jensen, told me he is open to the idea if OneWhale and its partners can secure the necessary permits and funds.

This tactical shift is one of several recent developments that strained the already fraught alliance between Crosby Haug and her scientific advisers. Moving Hvaldimir to Svalbard may well be his best chance of reuniting with wild belugas, but several experts I interviewed expressed serious concerns about the plan. Beluga societies tend to be highly dynamic and accepting—wild belugas have even adopted lone narwhals—but the Svalbard population is small, insular, and nonmigratory. Given Hvaldimir's mysterious origins and how much time he has spent away from his kin, there's no guarantee that the Svalbard belugas will welcome him, especially if he was caught in the distant Sea of Okhotsk, where the Russians reportedly acquire many of their military cetaceans. He might also introduce foreign diseases, pathogens, or unfavorable genetic mutations.

Moreover, Svalbard's remoteness and extreme weather make the expedition itself arduous and costly, not to mention stressful and disorienting for a beluga. Even if transport is successful, Hvaldimir would ideally require a period of acclimation on site before release, which would mean obtaining legal authorization to construct a temporary enclosure and maintaining it in potentially harsh conditions. In order to determine Hvaldimir's fate, scientists would have to secure tracking devices to his dorsal ridge with steel rods, a procedure that sometimes causes significant wounds and infections.

"I'm always in favor of getting animals into a more natural scenario, but you have to do it on a case-by-case basis," says Ingrid Visser, a whale scientist who is known for her studies of orcas and spent several weeks observing Hvaldimir. "It has to be driven by the welfare of the animal first and foremost, and it has to be backed by robust and compassionate science."

Last summer, following intense disagreement over Hvaldimir's future, a majority of OneWhale's scientific advisers, including Visser and every other cetacean expert, resigned. By September, Strand had left as well. Several of OneWhale's former members claim that the organization's leadership demonstrated a pattern of miscommunication, recklessness, and a disregard for scientific expertise. They say that Crosby Haug presented the Svalbard proposal to the Norwegian government without properly consulting them and that she did not clearly convey the regulatory hurdles to OneWhale's plans, namely the Norwegian laws that would complicate the confinement of a wild whale, even in the context of rehabilitation. (Crosby Haug and Siri Martinsen, leader of NOAH, dispute this.) They further contend that she spent too much time interacting with Hvaldimir in the water despite lacking the appropriate training, potentially reinforcing his dependency on humans and inadvertently encouraging tourists to do the same. (Crosby Haug denies this as well.)

The motives behind Crosby Haug's conduct became another significant point of contention. When Crosby Haug first traveled to Hammerfest, she did not intend to mix artistry with advocacy or to be a character in her own documentary. As her devotion to Hvaldimir deepened, however, she decided to explore their relationship on camera in a similar manner as the hit Netflix film *My Octopus Teacher*. Many of OneWhale's scientific consultants worried that Crosby Haug's attempt to steer Hvaldimir's future and simultaneously fulfill her cinematic aspirations created a conflict of interest. Over the past year, their latent uneasiness grew into distrust.

"OneWhale unraveled because of one person," says Stephen McCulloch, a longtime marine-mammal welfare specialist and former OneWhale adviser. "From my perspective, the problem was that you had a very controlling individual who had very little integrity or respect for a group of experts that really wanted to help. It became evident that Regina's priority was to make a film with a satisfying ending—belugas swimming off together in the sunset. But you can't film an animal that died because you didn't understand enough about him."

Crosby Haug denies prioritizing her film over Hvaldimir's welfare and claims that there is a proactive effort to disparage her. She says that McCulloch and others who left repeatedly proposed interventions that were unacceptable to OneWhale. "It is appropriate that he and others are no longer associated with us," she says, "because there are major differences in philosophy."

After leaving OneWhale, Strand founded his own organization, Marine Mind, to independently monitor and protect Hvaldimir. Most of the experts who resigned from OneWhale, including Visser, McCulloch, and Reiss, are now assisting Marine Mind, as are many of its former volunteers. Following these departures, OneWhale hired the marine biologist Anna Victoria Pyne Vinje as its new lead of science and research. The choice is controversial because Vinje, who strongly supports the Svalbard proposal, is also involved in a study on cetacean physiology that some conservationists have called "cruel and pointless." The ongoing research entails stretching nets across a known migration route in order to trap young minke whales and using superficial electrodes to test their hearing. Vinje and her collaborators maintain that the study is justified because it may ultimately improve efforts to protect baleen whales from noise pollution. But critics counter that the experiments are unnecessarily stressful to the animals and point out that a loose net has already entangled and drowned a passing whale.

This past fall, Hvaldimir reversed his southward trend and crossed back into Norway, where he is living around salmon farms and steadily regaining weight. Many of the experts I interviewed, while struggling to identify a clear solution to Hvaldimir's current situation, tentatively favored a kind of detached watchfulness, refraining from social interaction and intervening only in emergencies. In contrast, Reiss, a professor of psychology at Hunter College who has long studied cetacean cognition and communication, thinks that Hvaldimir is a strong candidate for a sanctuary because he is still too habituated to humans and thus exposed to undue risk. For his part, Strand says he is undertaking a research-and-development phase to determine the best strategy.

In the meantime, tensions between OneWhale and Marine Mind are becoming somewhat disruptive. Hvaldimir has recently interfered with or damaged propellers, sensors, and other equipment at the commercial salmon farms he frequents. In the past, employees on site were willing to deal directly with Strand to minimize Hvaldimir's mischief. Now, pressured by rival organizations whose members make conflicting statements, some farmers have instead referred him to their companies' public relations departments. "To be completely candid," Strand told me, "it's a mess."

When I began reporting this article, Hvaldimir had not yet commenced his southward odyssey. By the time I connected with OneWhale on the ground in late July, he was already in Sweden. I had told my editors that, although there were no guarantees, the likelihood of meeting Hvaldimir was high. I was nowhere near prepared for just how challenging the trip would be. Because Hvaldimir does not have any tracking devices, the task of finding him depends on social media posts and a tip line. We spent more than a week searching for him by car, train, and boat, hindered by strong winds and rough seas, often arriving at his last known location too late to encounter him. Exhausted, over budget, and faced with the prospect of even worse weather in coming days, I reluctantly accepted that it was time to return home. Just one hour before I planned to head to the airport, while I was still aboard OneWhale's rented catamaran, a report came in: someone had spotted Hvaldimir in a harbor on a small island about twenty miles northwest of Gothenburg. By a stroke of luck, we were only thirty minutes away.

When we arrived at the harbor, Hvaldimir was gliding just beneath the surface—an indistinct milky shape, noiseless and ghostlike. Close to two dozen spectators wandered the docks, trying to get a better look. Within seconds of our arrival, Hvaldimir swam toward us. Crosby Haug stood near the edge of our boat in a black wet suit, waving and calling out in that lilting, high-pitched voice reserved for pets and infants. "Hvaldimir! Hello!" she said, as the whale swam directly below us. "I'm coming, baby, I'm coming. Stay with us."

For the next hour, Hvaldimir followed our vessel and several others through the sheltered waters within the surrounding cluster of islands. At first, he remained largely underwater, breaching only momentarily to breathe. Gradually he began to bring his bulbous head above the surface, turning it from side to side as he inspected us with beady black eyes. His intelligence and curiosity were palpable. Whereas most whales and dolphins have fused neck bones and fixed expressions, belugas can flex their heads and alter the shape of their mouths, making them particularly expressive.

By the time Hvaldimir had returned to the central pier, a substantially larger crowd had gathered. Children and adults alike thronged the docks, dangling feet and hands in the water, hoping to touch the celebrity cetacean or at least get a photo. Hvaldimir was docile and playful, swimming right up against people's shins, allowing them to pet his head and back and repeatedly offering the underside of his flipper for a high five—one of many tricks he presumably learned in captivity.

Crosby Haug pulled her blond curls into a ponytail and entered the water, swimming alongside Hvaldimir and explaining how to interact with him safely. "It's okay to touch him," she said at one point, "just not his eyes or his blowhole." When other people tried to get in the water, she cautioned them. "We're with his science team," she said, "and it's not recommended to swim with him. I'm just letting you know that. Today we're seeing how much fat he has. You see this shallow part underneath his blowhole? That means he is losing too much weight. There's not enough fish in the water down here. So we're trying to get him back up to Norway."

A little while later, Crosby Haug enticed Hvaldimir to stay near the catamaran, away from the crowd. She played with him for about an hour, throwing small white buoys for him to fetch and allowing him to nudge her around. Eventually she decided it was time to leave the harbor and encourage Hvaldimir to follow the catamaran north. We loosened the ropes tying the boat to the docks and motored away. Hvaldimir swam alongside us, undulating tirelessly just beneath the surface, the hum of the engines punctuated by his powerful exhalations. The farther we traveled, however, the more difficult it became to keep Hvaldimir at our side. He was easily distracted, veering away to inspect other vessels. Frustrated by repeated interruptions, Crosby Haug started to yell at passing boats, waving her arms and warning them to stay away.

Like many celebrities, Hvaldimir has lived a life defined by other people's desires. Almost everyone he has met wants something from him: a snapshot, a story, a lifetime of submission. One of the most tragic aspects of his predicament is the discrepancy between how much he is adored and how little has been accomplished to secure his long-term welfare. Hvaldimir ostensibly offers our species a chance at redemption: a formerly captive whale, already moving freely through the ocean, requiring only some redirection to reunite with his kind. But the enormity of what we have done to him and so many other sentient beings like him severely complicates—and in some cases prohibits—such a reversal. Hvaldimir is so far displaced from his origins—geographically, ecologically, culturally—that it's not clear whether a homecoming is still achievable.

From ocher bison painted on cave walls to the elephants in Europe's medieval menageries to ongoing killer-whale shows and interactive dolphin pools, humans have long been enamored with other large, social, and intelligent animals. We love them because they are simultaneously familiar and exotic—because they both mirror us and represent ways of being beyond our ken. We have often expressed our passion for such creatures by trying to possess them: by fitting them with collars, roping them into circuses, and placing them behind glass. Even the military conscription of marine mammals is a kind of admiration, or at least recognition, of their extraordinary abilities. Yet the closer we have pulled such animals toward us, the more difficult it has become to deny the torment that our proximity inflicts. Perhaps the purest act of love is to leave them alone in the first place.

After traveling about eight miles northwest of the harbor where we first found Hvaldimir, he began to slow down and trail off more frequently, possibly losing interest, stamina, or both. As we approached a town called Skarhamn, he vanished amid choppy water. Studying the ocean surface in the quickly fading light, you could easily mistake the white crest of a wave or a patch of foam for a dorsal ridge or fluke. On a hunch, we searched a nearby harbor, where we glimpsed Hvaldimir tugging a ship's ropes. Seconds later, he slipped beneath the slate-blue sea. With little recourse in the dark, we found a place to dock, hoping, rather helplessly, that the world's most famous beluga—the half-wild whale we had chased for more than a week and who halfheartedly chased us back for all of an hour—might decide to stay with us through the night. The next morning, he was nowhere to be seen.

DHRUV KHULLAR

No Time to Die

FROM *The New Yorker*

SOME OF MY earliest memories are of summers with my grandparents, in New Delhi. I spent long, scorching months drinking lassi, playing cricket, and helping my grandparents find ripe mangoes at roadside markets. Then I'd return to the U.S., my English rusty from disuse, and go months or years without seeing them. At some point, my India trips started to feel like snapshots of loss. My grandfathers died suddenly, probably of heart attacks. My Biji, my father's mother, fell and broke her hip in her seventies, and she spent her last years moving back and forth between her bed and her couch. My Nani, my mother's mother, developed excruciating arthritis in both knees; in order for her to leave her fifth-floor walk-up, my uncle practically had to carry her down the stairs. I have always wondered whether their fading vitality—the way their worlds contracted and their possibilities vanished—was an inevitability of aging or something that could have been averted.

Many of us have come to expect that our bodies and minds will deteriorate in our final years—that we may die feeble, either dependent or alone. Paradoxically, this outcome is a kind of success. For most of history, humans didn't live long enough to confront the ailments of old age. In 1900, a baby born in the U.S. could expect to live just forty-seven years, and one in five died before the age of ten. But twentieth-century victories against infectious diseases—in the form of sanitation, antibiotics, and vaccines—dramatically extended lifespans, and today the average newborn lives to around seventy-seven. Lately, though, progress has slowed. In the past six decades, medicine has added about seven years to the average lifespan—less by saving young lives than by extending old ones, and often in states of ill health. In many cases, we're prolonging the time it takes to die.

A growing number of celebrity doctors, futurists, and so-called biohackers now argue that it doesn't have to be this way. There are, by some estimates, hundreds of specialized "longevity clinics"—including some that charge six-figure annual fees—which claim to offer more of the world's most valuable commodity: years of healthy life. Perhaps the most prominent longevity evangelist is Peter Attia, the author, with Bill Gifford, of the best-selling book *Outlive: The Science and Art of Longevity*. Through his telemedicine practice in Austin, Texas, for an undisclosed price, Attia offers health advice, diagnostic tests, exercise protocols, and supplements to a wealthy and exclusive clientele. He also interviews an eclectic mix of scientists, doctors, and entrepreneurs for a popular podcast, "The Drive." Oprah has interviewed him; Hugh Jackman and Gwyneth Paltrow follow his regimens.

Attia graduated from medical school and trained to be a surgeon, but grew disillusioned during residency and dropped out. He became a consultant for McKinsey instead, and then worked for an energy company. Finally, in his mid-thirties, a fixation with his own health brought him back to medicine. As a new father, he learned that he was prediabetic, and he reflected on men in his family who'd died early, of heart disease. In his book, he describes his former physique as "sausage-like"; on a beach one day, his wife told him, "Peter, I think you should work on being a little less not thin." Soon, he was "down the rabbit hole of complete physical optimization."

Attia, now fifty-one, has become convinced that science, technology, and targeted work can solve a uniquely modern problem: the "marginal decade" at the end of our lives, when medicine keeps us alive but our independence and capacities bleed away. It's a scandal, in his view, that our lifespan has grown so much more than our *health* span. Many of Attia's prescriptions are obvious: work out, eat healthily, sleep well, nurture relationships. (The Harvard Study of Adult Development has found, in eight decades, that human connections may be the single most critical determinant of long-term happiness and health.) But Attia often extrapolates from scientific data to offer jarringly intense and specific advice. Want to be able to lift a grandchild when you're eighty? Goblet squat fifty-five pounds when you're forty. Hope to lift yourself off the floor unassisted in old age? Try "toe yoga." Attia notes that each decade after thirty we lose a meaningful amount of muscle mass and cardiovascular fitness. If we wish to slow that decline, and to complete a "Centenarian Decathlon" of important late-in-life activities—carry groceries, climb stairs, have sex—we need to become "athletes of life."

The increasing obsession with longevity has inspired a backlash. Many in the life-extension movement are quacks or hacks who peddle pills, potions, and false promises; longevity skeptics tend to see the loss of our capacities as something to accept, not avoid. Ezekiel Emanuel, an oncologist and a health-policy professor at the University of Pennsylvania, derides Attia as an "American immortal" who overcomplicates straightforward advice. "The idea that you're going to get another healthy decade of life just by doing the things he says is hocus-pocus," Emanuel, who served as a special adviser to the Obama administration, told me. "No one's got that evidence." Half an hour of daily exercise clearly improves and extends lives, but it's hard to prove that Attia's intensive regimens are much more beneficial. By incessantly preparing for the future, the skeptics say, we mistake a long life for a worthwhile one.

On a recent afternoon, I chased my three-year-old daughter around the playground for an hour. When we returned home, she spread a jigsaw puzzle out on the floor and looked up expectantly. I liked the idea of sitting still, but my knees hurt and my back was tense. I had to transfer the puzzle to a grown-up table and sit my daughter in a booster seat. She didn't seem to mind, but I remember that day as the first time that my physical limits noticeably constrained what we could do together. Longevity has become a concrete problem, just as it was for my grandparents: I wake up with aches in long-ignored joints and tendons; I calculate, with dismay, the age I'll be when my children graduate from college or start their own families. One day, we're going to die. What should that mean for how we live today?

In 1980, James Fries, a Stanford rheumatologist, predicted in *The New England Journal of Medicine* that better medicines and behaviors would soon enable a "compression of morbidity," which would delay disease and debility until the very end of our lives. In the late '90s, Fries supported his hypothesis by publishing a decades-long study. University of Pennsylvania graduates who, in their forties, exercised more, weighed less, and didn't smoke much were half as likely as others to suffer a significant disability in their seventies; they seemed to postpone the onset of disability by more than five years. But the alumni of an elite college may not have been representative. Fries died of end-stage dementia in 2021, at the age of eighty-three, and his broader prediction never seemed to come true. If anything, longer lives now appear to include *more* difficult years. The "compression of morbidity may be as illusory as immortality," two demographers, Eileen Crimmins and Hiram Beltrán-Sánchez, wrote in 2010. According to the World Health Organization, the average American can expect just one healthy birthday after the age of sixty-five. (Health spans are greater in countries such as Switzerland, Japan, Panama, Turkey, and Sri Lanka.)

Earlier this year, I flew to Austin to hear Attia's thoughts about how to change that. When my Uber pulled into his driveway, he was finishing his morning workout, so an assistant led me inside, where I spent a few moments looking at Formula 1 paraphernalia. Floor-to-ceiling windows overlooked lush hills, and a herd of life-size elk and deer models stood in the sun. I would soon learn that they served as targets for archery practice.

Attia has a shaved head, a sharp nose, and a stubbled chin, which make him look like a mix of Stanley Tucci and Jeff Bezos. When he appeared, he was wearing a fitted T-shirt that emphasized his biceps. He led me to his kitchen, where he offered me coffee and mixed a brightly colored concoction for himself. "I always try to get some protein in the morning," he said. He eats as many as six sticks of venison jerky a day.

When I asked Attia about the longevity movement, he bristled. The term "just smells of snake oil," he said. "Most of what I see out there is what I think of as sci-fi longevity. *We're going to live to be two hundred, and death is going to become irrelevant.*" He handed me my coffee. "The way I talk about it is in a very low-tech way." Attia has said that he wouldn't want to live forever, even if he could, and he seems wary of a stereotype of longevity doctors. At parties, he sometimes pretends he's a race-car driver or a shepherd. "I thought I was going to get skewered for writing *Outlive*," he told me. "I thought doctors were going to say, 'This guy is a grifter. He doesn't know what he's talking about.'" Some do say that—but others have become his followers.

Like a consultant, Attia often explains himself using frameworks. In the time of Hippocrates, he said, there was Medicine 1.0, a pre-modern system of diagnosis and treatment based on observation, anecdote, and guesswork. In the twentieth century, Medicine 2.0 deployed the randomized controlled trial to produce scientific marvels such as dialysis, organ transplants, and antiviral drugs. But Attia also sees Medicine 2.0, arguably the type of medicine that I practice, as passive. It often acts after the onset of damage—debilitating arthritis, a broken hip—instead of aggressively and proactively warding off illness and injury. Attia preaches Medicine 3.0.

We sat down at a sort of command center in his home office. On a large monitor, Attia pulled up a patient's "longevity risk assessment"—his team's calculation of threats to life and limb, ranked by relative importance both now and in the future. Although Attia describes his approach as low-tech, his patients receive dozens of tests, some of them outside the medical mainstream: full-body MRIs, body-fat-composition scans, DNA analyses. He often screens for Alzheimer's risk, something that many doctors advise against, in part because patients can't do much about a distressing result. (While filming an episode of the longevity docuseries *Limitless*, Chris Hemsworth, the host, learned from Attia that he has a gene associated with a roughly eight-hundred-percent increase in Alzheimer's risk; afterward, Hemsworth took time off from acting.)

The assessment on Attia's monitor was for a middle-aged patient who had been given a diagnosis of attention-deficit disorder, and had also undergone several surgeries. Running down the left side of a chart was a list of conditions, starting with the ones that posed the greatest risk: emotional-health problems and physical injury. The right side showed percentages—the guesstimated chance that each condition would become an issue in the future. Cancer and neurodegenerative disease were small risks now, but they ballooned into dominant hazards later in life. "It's more art than science," Attia told me. "There's no AI that's ever going to be able to spit out these numbers. It requires clinical judgment."

The patient's results seemed to trouble Attia. "They're on a path to have a very physically debilitated last decade of life," he said. (His practice omitted the names of patients and asked me to change their pronouns, to protect their privacy.) Attia's staff focused on three near-term suggestions: see a recommended therapist, work with a recommended exercise team, get a colonoscopy. They may have been powerful in their simplicity, but they were standard enough that I wondered whether all that testing was really necessary.

Many more patients will soon receive a version of Attia's advice. Not long ago, he debuted a program called Early, a kind of MasterClass on longevity that costs twenty-five hundred dollars to access online. In a series of slick videos, Attia—in the hybrid persona of doctor, teacher, and coach—sits in a leather chair and talks about anticipating and averting diseases. Soft music plays; one video cuts to Attia's muscles tensing as he aims a manual crossbow and strikes a faraway target. "Hope is not a strategy," he says later, looking into the camera. "The marginal decade is the most important decade."

When I asked Attia what he sees as a threat to his own longevity, he spoke of a lifelong battle to keep his emotions in check. He previously described himself as a workaholic with anger-management issues. During a business trip in 2017, he received a call from his wife, who was in a state of terror: their month-old son had stopped breathing and turned blue, and didn't have a pulse. She'd started CPR on his tiny body. By the time first responders arrived, he'd regained a heartbeat and his skin was transitioning to pink.

"Okay, call me when you get to the hospital, so I can talk to the doctors in the ICU," Attia remembers telling her, in his book. Then he went to dinner. Ten days passed before he flew home.

"I don't try to forgive myself for it," Attia told me recently. "There was a really, really broken person who did those things." He paused, as though replaying the episode in his mind. "Don't ever forget that that bastard is out there. . . . And make sure you work really hard to keep him at bay."

The quest for physical optimization can easily become a substitute for deeper fulfillment. A decade ago, Attia exercised twenty-eight hours a week and observed a strict ketogenic diet. His "biomarkers were out of this world," he has said, but he refused cookies that his children baked for him and pasta during trips to Italy. "I was doing *everything* to live longer, despite being completely miserable emotionally," he writes in *Outlive*. In a recent interview with the *Times*, Attia said that, before attending an event at his son's kindergarten, he thought for a moment of the downsides: it would eat into his time for squats and deadlifts. "That's costing me a little in terms of fitness," he said. "But that's the trade-off I wanted to make."

Attia wouldn't tell me how many patients he sees, or how much they pay; there has been speculation that his services cost around a hundred and fifty thousand dollars a year. One of his clients, Carl Barney, is the eighty-two-year-old founder of a nonprofit inspired by Ayn Rand. He told me that, during the past three years, Attia has encouraged him to diversify his exercise routines and sleep more. At Attia's urging, he now scoops collagen protein powder into his morning tea, drinks bone broth before dinner, and tries to consume another hundred-plus grams of protein during the day. I asked him whether Attia's advice was worth the high price. "I have wealth, so, for me, it's a bargain," he said.

Attia let me join a video call in which five telegenic members of his team discussed new patients. A primary care doctor told the group about a patient who managed career stress by engaging in extreme sports. "If they can't do those things, a lot of their coping mechanisms will crumble," she said. But it is difficult to quantify the risk that an injury might incapacitate someone. "We have questions like 'Do you wear a seat belt?'" the doctor said. "We don't have questions like 'Do you cliff-jump into the ocean?'"

The patient's tests had identified hearing and vision issues; a full-body MRI showed ambiguous cysts in an internal organ. Genetic screenings suggested early-dementia risks, and the patient had recently asked the team, "Should I just quit my job right now and focus on living my life?" But a neurologist observed that the patient's relatives had developed dementia only later in life. "This is probably going to be the most hopeful piece of news we give them," Attia said. "We can say, clearly, no. . . . We have a lot of runway."

I was envious that the doctors could pay so much attention to one patient. Attia had time to ask how well this person flossed—something I don't ask my wife, let alone the patients I see in fifteen-minute increments. But the primary care doctor made me ponder whether there was such a thing as too much attention. The patient struggled with anxiety, she said, and seemed to be looking for validation of their fitness routine. "Instead, we show them all these medical problems," she joked. "And then we're, like, 'Why are they so anxious?'"

A few years ago, Attia started an intense fasting regimen. Some lab studies have suggested that fasting may reduce inflammation and clear the body of precancerous cells, but there is no agreed-upon "dose" for healthy fasting; Attia decided that for about two weeks each quarter he would drink only water. After losing fifteen pounds of muscle, however, he abandoned the practice. "He has always been extreme," Steven Levitt, the University of Chicago economist and co-author of *Freakonomics*, who is a patient and a friend of Attia's, told me. "Peter has been wrong a lot, but he changes his view when he runs into evidence that contradicts his theory." Levitt trusts and admires Attia, but acknowledged that Attia's fans in the longevity movement could go too far. "Followers often become more extreme than the leaders," he told me.

Attia takes rapamycin, a drug that affects the immune system and is normally used by organ-transplant patients; the drug, which occasionally gives him mouth sores, has seemed to lengthen the lives of laboratory animals. Animal experiments are unreliable indicators of how drugs work in humans, and rapamycin is not endorsed by the Food and Drug Administration or medical societies for prolonging life. "This isn't like a vitamin," Eric Topol, a cardiologist and the director of the Scripps Research Translational Institute, told me. "It's a serious drug that can potentially lower your body's ability to fight off infections." When I asked Attia why he has promoted the drug in his book and on his podcast, he said, "There's a big difference between saying everyone should take rapamycin and saying, 'I take rapamycin.'" He also cited a study suggesting that rapamycin bolstered immunity in elderly people.

It's clear that Attia *has* prompted people to take the drug. Topol said that some of his patients had read *Outlive* and then asked him for prescriptions. "People see him as the expert, so they are going to try something if he says he's doing it," he told me. "His followers aren't going to be able to detect which recommendations are firmly grounded in evidence." Meanwhile, training dozens of hours a week might take more time than it will ever tack on; good health could even drag out some terminal illnesses. "Peter's theory of Medicine 3.0 is that you get this long life where you're healthy, and then you fall off a cliff," Topol said. "It would be great if it were true. There isn't any evidence for it."

Emanuel, the University of Pennsylvania professor, has said that he wants to live to seventy-five. (He is sixty-six.) "Living a long time is not an end in itself," he told me over Zoom. "If it becomes the focus of your life . . . that is one of the worst mistakes you can make." It's not that we *shouldn't* exercise or eat well— but "everyone goes through a decline," Emanuel said. "Spending your life worried about all these tiny things is a waste of time."

During our video call, Emanuel walked to his bookshelf, pulled out a copy of *Outlive*, and read a line from the epilogue: "'It was only after much reflection on this whole experience that I really

began to understand how emotional health relates to longevity.'"
He slapped his palm against his forehead in a mock epiphany.
"Oh, really? Longevity doesn't matter if your life sucks? . . . I
mean, come on. It's ridiculous."

Leon Kass, who served as the chair of the President's Council
on Bioethics under George W. Bush, has written that losing our
capacities might be a kind of prerequisite to accepting our mortality: maybe the slowing of body and mind is what makes death
tolerable. He quotes Michel de Montaigne, the sixteenth-century
essayist. "Inasmuch as I no longer cling so hard to the good
things of life when I begin to lose the use and pleasure of them,
I come to view death with much less frightened eyes," Montaigne
wrote. "When we are led by Nature's hand down a gentle and virtually imperceptible slope, bit by bit, one step at a time, she rolls
us into this wretched state and makes us familiar with it."

Nir Barzilai, the director of the Institute for Aging Research at
the Albert Einstein College of Medicine, is sometimes credited
with discovering the first "longevity gene," an unusual variation
in DNA that is linked to exceptionally long life. When people
ask him to define aging, however, he doesn't talk about that, and
instead tells a two-sentence story. An elderly woman turns to her
husband and says, "Honey, why don't we go upstairs and make
love?" Her husband responds, "Sweetie, I cannot do both."

For decades, Barzilai has followed hundreds of Ashkenazi
Jews and examined genetic, behavioral, and environmental fac-
tors that have helped them age past ninety-five in exceptional
health. Centenarians die of the same things as the rest of us,
but later, he told me. You might live to a hundred if you could
pick your genes—picking the healthy lunch option isn't likely to
be enough. Strikingly, around half the centenarians in Barzilai's
study have been overweight; 30 percent of the women and 60
percent of the men have been heavy smokers. When he asked
a hundred-year-old whether anyone had warned her about the
harms of tobacco, she responded, "Yes, all four of the doctors
who told me to stop smoking—they died."

Attia tends to argue that individual choices matter not because they are all-powerful but because they are the power that

we have. He compares healthy aging to investing in retirement: contribute what you can, whether it's a daily walk or an extra half hour of sleep, and the benefits may compound over time. "If I'm being brutally honest, I think some people are looking for a reason not to do it because it's hard," he said.

He acknowledged that health, like wealth, is unequally distributed; indeed, one of the most powerful longevity "medicines" is money, which can buy people less stress, better education, safer neighborhoods, and higher-quality medical care. For this reason, Emanuel argues that doctors should focus less on "getting rich people from ninety to a hundred" than on improving health in communities where people die young. When I asked Attia whether his practice might perpetuate this divide, he told me, "I've never concerned myself with that. I don't think it's unimportant, but it's absolutely not a problem I'm interested in addressing." He pointed out that his podcast is free and his book costs less than twenty dollars.

Attia has a term for his unproven ideas: evidence-informed medicine, or interventions that rest more on theory than on randomized controlled trials in humans. In his view, medical recommendations are often too conservative; in some cases, rigorous studies would not only take too long but also be unethical or impractical. (Try randomly assigning babies to ketogenic diets at birth.) According to Attia, medicine's admonition to do no harm is "sanctimonious bullshit" that steers doctors toward passivity and resignation. But this line of reasoning leaves enormous room for extrapolation, and could be used to justify almost any practice. Talking to Attia, I frequently reminded myself that I can't diet and exercise my way out of many diseases, and there's no regimen to eradicate uncertainty. Still, I felt the pull of becoming an "athlete of life." Too often, conversations about life extension devolve into unhelpful abstractions and untestable speculation; one appeal of Attia's advice is that it's so tangible. Critics can paint his counsel as blindingly obvious or needlessly complex—but he has, at least, inspired large audiences to imagine what a better approach to aging could look like. "There is actually no such thing as atheism," David Foster Wallace once said. "The only choice we get is what to worship." In a society

that chases money, power, fame, and beauty, there are worse gods
than longevity.

Attia is a history buff and an avid rucker, which means that he
likes to lug a heavy rucksack on long walks. In June, to mark the
eightieth anniversary of D Day, Attia plans to set out from Utah
Beach, in Normandy, with four friends, carrying supplies plus a
twenty-five-pound weight, just because. They hope to trek eighty
kilometers through the night and arrive at Omaha Beach less
than twenty hours later.

During my visit, Attia agreed to take me on a gentler ruck,
in the hills around his neighborhood. We had lunch delivered
beforehand; he had a salad with chicken, nuts, and balsamic vin-
egar, while I had a saucy pasta. I caught him glancing at my bowl
and imagined him judging my macronutrients. Then I changed
into my workout gear and met him in his garage, where he outfitted me with an army-green rucksack. "I generally recommend
that people new to rucking start light," he said, and slipped a
thirty-pound weight plate into my bag. He added a few to his
own, and we set out into the afternoon sun.

After climbing a hill, we entered a clearing. Around us, a circle of oak trees stretched majestically into the air. A mild breeze
cooled the sweat on my brow, and a flock of birds darted across
the open sky. I considered stopping to look around, but I could
hear the weights in Attia's rucksack clanging like a metronome,
so I sped up. I was feeling pretty good; I made a mental note to
offer to take one of his plates. I changed my mind after the next
hill, when my neck stiffened and my shoulders started to hurt.

"Why not just go for a hike?" I asked, panting.

"I wouldn't find that as fun," he said. "Plus, it's not as good for
your trunk."

In the distance, a school bus drove by. I envisioned myself lifting my hypothetical granddaughter one day, then adjusted my
rucksack and straightened my posture.

As we walked on, I thought about a curious body of psychological research, which suggests that as we age and lose our capacities we tend to grow more content, not less. This finding
clashes with popular conceptions of getting older, but seems to

hold across continents, cultures, and eras. "I can't do everything I used to," a family friend, who is in his eighties and has been married for sixty years, recently told me. "But I wouldn't say I'm any less happy than I was before." Lost pleasures, he said, could sometimes be replaced: rounds of golf gave way to brisk walks, and when walking became difficult he spent more time talking to his children and grandchildren. As we grasp that our days are limited, we seem to abdicate our need for control; we may try to close the gap between what we want and what we have. Healthy aging seems to require a shift in mindset as much as a shift in muscle mass.

My calves started to burn. I felt a knot in my back. White clouds veiled the sun, and a few ethereal rays came through. It looked so much like the entrance to TV Heaven that I half expected a deep voice to boom from above.

"Sometimes I think about all the people who've ever lived, and how lucky I am to be alive right now," Attia told me. "Like, if I died tomorrow, it would be okay." We started down a final hill. His house came into view. "But, while I'm here, I want to know that I gave it my all," he went on. "We have this one shot. Wouldn't it be a shame if we didn't make the most of it?"

SHARON LERNER

Toxic Gaslighting: How 3M Executives Convinced a Scientist the Forever Chemicals She Found in Human Blood Were Safe

FROM *The New Yorker / ProPublica*

KRIS HANSEN HAD worked as a chemist at the 3M Corporation for about a year when her boss, an affable senior scientist named Jim Johnson, gave her a strange assignment. 3M had invented Scotch Tape and Post-it notes; it sold everything from sandpaper to kitchen sponges. But on this day, in 1997, Johnson wanted Hansen to test human blood for chemical contamination.

Several of 3M's most successful products contained man-made compounds called fluorochemicals. In a spray called Scotch-gard, fluorochemicals protected leather and fabric from stains. In a coating known as Scotchban, they prevented food packaging from getting soggy. In a soapy foam used by firefighters, they helped extinguish jet-fuel fires. Johnson explained to Hansen that one of the company's fluorochemicals, PFOS—short for per-fluorooctanesulfonic acid—often found its way into the bodies of

3M factory workers. Although he said that they were unharmed, he had recently hired an outside lab to measure the levels in their blood. The lab had just reported something odd, however. For the sake of comparison, it had tested blood samples from the American Red Cross, which came from the general population and should have been free of fluorochemicals. Instead, it kept finding a contaminant in the blood.

Johnson asked Hansen to figure out whether the lab had made a mistake. Detecting trace levels of chemicals was her specialty: She had recently written a doctoral dissertation about tiny particles in the atmosphere. Hansen's team of lab technicians and junior scientists fetched a blood sample from a lab-supply company and prepped it for analysis. Then Hansen switched on an oven-size box known as a mass spectrometer, which weighs molecules so that scientists can identify them.

As the lab equipment hummed around her, Hansen loaded a sample into the machine. A graph appeared on the mass spectrometer's display; it suggested that there was a compound in the blood that could be PFOS. That's weird, Hansen thought. Why would a chemical produced by 3M show up in people who had never worked for the company?

Hansen didn't want to share her results until she was certain that they were correct, so she and her team spent several weeks analyzing more blood, often in time-consuming overnight tests. All the samples appeared to be contaminated. When Hansen used a more precise method, liquid chromatography, the results left little doubt that the chemical in the Red Cross blood was PFOS.

Hansen now felt obligated to update her boss. Johnson was a towering, bearded man, and she liked him: He seemed to trust her expertise, and he found something to laugh about in most conversations. But, when she shared her findings, his response was cryptic. "This changes everything," he said. Before she could ask him what he meant, he went into his office and closed the door.

This was not the first time that Hansen had found a chemical where it didn't belong. A wiry woman who grew up skiing competitively, Hansen had always liked to spend time outdoors; for her chemistry thesis at Williams College, she had kayaked around the former site of an electric company on the Hoosic River, collecting crayfish and testing them for industrial pollutants called polychlorinated biphenyls, or PCBs. Her research, which showed that a drainage ditch at the site was leaking the chemicals, prompted a news story and contributed to a cleanup effort overseen by the Massachusetts Department of Environmental Protection. At 3M, Hansen assumed that her bosses would respond to her findings with the same kind of diligence and care.

Hansen stayed near Johnson's office for the rest of the day, anxiously waiting for him to react to her research. He never did. In the days that followed, Hansen sensed that Johnson had notified some of his superiors. She remembers his boss, Dale Bacon, a paunchy fellow with gray hair, stopping by her desk and suggesting that she had made a mistake. "I don't think so," she told him. In subsequent weeks, Hansen and her team ordered fresh blood samples from every supplier that 3M worked with. Each of the samples tested positive for PFOS.

In the middle of this testing, Johnson suddenly announced that he would be taking early retirement. After he packed up his office and left, Hansen felt adrift. She was so new to corporate life that her office clothes—pleated pants and dress shirts—still felt like a costume. Johnson had always guided her research, and he hadn't told Hansen what she should do next. She reminded herself of what he had said—that the chemical wasn't harmful in factory workers. But she couldn't be sure that it was harm*less*. She knew that PCBs, for example, were mass-produced for years before studies showed that they accumulate in the food chain and cause a range of health issues, including damage to the brain. The most reliable way to gauge the safety of chemicals is to study them over time, in animals and, if possible, in humans.

What Hansen didn't know was that 3M had already conducted animal studies—two decades earlier. They had shown PFOS to be toxic, yet the results remained secret, even to many at the company. In one early experiment, conducted in the late '70s, a group of 3M scientists fed PFOS to rats on a daily basis. Starting at the second-lowest dose that the scientists tested, about 10 milligrams for every kilogram of body weight, the rats showed signs of possible harm to their livers, and half of them died. At higher doses, every rat died. Soon afterward, 3M scientists found that a relatively low daily dose, 4.5 milligrams for every kilogram of body weight, could kill a monkey within weeks. (Based on this result, the chemical would currently fall into the highest of five toxicity levels recognized by the United Nations.) This daily dose of PFOS was orders of magnitude greater than the amount that the average person would ingest, but it was still relatively low—roughly comparable to the dose of aspirin in a standard tablet.

In 1979, an internal company report deemed PFOS "certainly more toxic than anticipated" and recommended longer-term studies. That year, 3M executives flew to San Francisco to consult Harold Hodge, a respected toxicologist. They told Hodge only part of what they knew: that PFOS had sickened and even killed laboratory animals and had caused liver abnormalities in factory workers. According to a 3M document that was marked "CONFIDENTIAL," Hodge urged the executives to study whether the company's fluorochemicals caused reproductive issues or cancer. After reviewing more data, he told one of them to find out whether the chemicals were present "in man," and he added, "If the levels are high and widespread and the half-life is long, we could have a serious problem." Yet Hodge's warning was omitted from official meeting notes, and the company's fluorochemical production increased over time.

Hansen's bosses never told her that PFOS was toxic. In the weeks after Johnson left 3M, however, she felt that she was under a new level of scrutiny. One of her superiors suggested that her equipment might be contaminated, so she cleaned the mass spectrometer and then the entire lab. Her results didn't change. Another encouraged her to repeatedly analyze her syringes, bags, and test tubes, in case they had tainted the blood. (They had not.) Her managers were less concerned about PFOS, it seemed to Hansen, than about the chance that she was wrong.

Sometimes Hansen doubted herself. She was twenty-eight and had only recently earned her PhD. But she continued her experiments, if only to respond to the questions of her managers.

3M bought three additional mass spectrometers, which each cost more than a car, and Hansen used them to test more blood samples. In late 1997, her new boss, Bacon, even had her fly out to the company that manufactured the machines, so that she could repeat her tests there. She studied the blood of hundreds of people from more than a dozen blood banks in various states. Each sample contained PFOS. The chemical seemed to be everywhere.

When 3M was founded, in 1902, it was known as the Minnesota Mining and Manufacturing Company. After its mining operations flopped, the company pivoted to sandpaper and then to a series of clever inventions aimed at improving everyday life. An early employee noticed that autoworkers were struggling to paint two-tone cars, which were popular at the time; he eventually invented masking tape, using crêpe paper and cabinetmaker's glue. Another 3M employee created Post-it notes to help him bookmark passages in his church hymnal. An official history of 3M, published for the company's hundreth anniversary, celebrated its "tolerance for tinkerers."

Fluorochemicals had their origins in the American effort to build the atomic bomb. During the Second World War, scientists for the Manhattan Project developed one of the first safe processes for bonding carbon to fluorine, a dangerously reactive element that experts had nicknamed "the wildest hellcat" of chemistry. After the war, 3M hired some Manhattan Project chemists and began mass-producing chains of carbon atoms bonded to fluorine atoms. The resulting chemicals proved to be astonishingly versatile, in part because they resist oil, water, and heat. They are also incredibly long-lasting, earning them the moniker "forever chemicals."

In the early '50s, 3M began selling one of its fluorochemicals, PFOA, to the chemical company DuPont for use in Teflon. Then, a couple of years later, a dollop of fluorochemical goo landed on a 3M employee's tennis shoe, where it proved impervious to stains and impossible to wipe off. 3M now had the idea for Scotchgard and Scotchban. By the time Hansen was in elementary school, in the '70s, both products were ubiquitous.

Restaurants served French fries in Scotchban-treated packaging. Hansen's mother sprayed Scotchgard on the living-room couch.

Hansen grew up in Lake Elmo, Minnesota, not far from 3M's headquarters. Her father was one of the company's star engineers and was even inducted into its hall of fame in 1979; he had helped to create Scotch-Brite scouring pads and Coban wrap, a soft alternative to sticky bandages. Once, he molded some fibers into cups, thinking that they might make a good bra. They turned out to be miserably uncomfortable, so he and his colleagues placed them over their mouths, giving the company the inspiration for its signature N95 mask.

Hansen never intended to follow her father to the company. She spent her childhood summers catching turtles and leopard frogs at the lake and hoped to have a career in environmental conservation. Her first job after earning her chemistry PhD was on a boat, which took her to remote parts of the Pacific Ocean. But the voyage left her so seasick that she lost twenty pounds, and she soon retreated to Minnesota. In 1996, at her father's suggestion, Hansen applied for a position in 3M's environmental lab.

After Hansen started her PFOS research, her relationships with some colleagues seemed to deteriorate. One afternoon in 1998, a trim 3M epidemiologist named Geary Olsen arrived with several vials of blood and asked her to test them. The next morning, she read the results to him and several colleagues—positive for PFOS. As Hansen remembers it, Olsen looked triumphant. "Those samples came from my horse," he said—and his horse certainly wasn't eating at McDonald's or trotting on Scotchgarded carpets. Hansen felt that he was trying to humiliate her. (Olsen did not respond to requests for comment.) What Hansen wanted to know was how PFOS was making its way into animals.

She found an answer in data from lab rats, which also appeared to have fluorochemicals in their blood. Rats that had more fish meal in their diets, she discovered, tended to have higher levels of PFOS, suggesting that the chemical had spread through the food chain and perhaps through water. In male lab rats, PFOS levels rose with age, indicating that the chemical accumulated in the body. But, curiously, in female rats the levels sometimes fell.

Hansen was unsettled when toxicology reports indicated why: Mother rats seemed to be offloading the chemical to their pups. Exposure to PFOS could begin before birth.

Another study confirmed that Scotchban and Scotchgard were sources of the chemical. PFOS wasn't an official ingredient in either product, but both contained other fluorochemicals that, the study showed, broke down into PFOS in the bodies of lab rats. Hansen and her team ultimately found PFOS in eagles, chickens, rabbits, cows, pigs, and other animals. They also found fourteen additional fluorochemicals in human blood, including several produced by 3M. Some were present in wastewater from a 3M factory.

At one point, Hansen told her father, Paul, that she was frustrated by the way senior colleagues kept questioning her work. Paul had recently retired, but he had confidence in 3M's top executives, and he suggested that she take her findings directly to them. But as a relatively new employee—and one of the few women scientists at a company of about seventy-five thousand people—Hansen found the idea preposterous. When Paul offered to talk to some of 3M's executives himself, she was mortified at the idea of her father interceding.

Hansen knew that if she could find a blood sample that *didn't* contain PFOS then she might be able to convince her colleagues that the other samples did. She and her team began to study historical blood from the early decades of PFOS production. They soon found the chemical in blood from a 1969–71 Michigan breast cancer study. Then they ran an overnight test on blood that had been collected in rural China during the '80s and '90s. If any place were PFOS-free, she figured, it would be somewhere remote, where 3M products weren't in widespread use.

The next morning, anxious to see the results, Hansen arrived at the lab before anyone else. For the first time since she had begun testing blood, some of the samples showed no trace of PFOS. She was so struck that she called her husband. There was nothing wrong with her equipment or methodology; PFOS, a man-made chemical produced by her employer, really *was* in human blood, practically everywhere. Hansen's team found it in Swedish blood samples from 1957 and 1971. After that, her lab analyzed blood that had been collected before 3M created PFOS. It tested negative. Apparently, fluorochemicals had entered human blood after the company started selling products that contained them. They had leached out of 3M's sprays, coatings, and factories—and into all of us.

That summer, an in-house librarian at 3M delivered a surprising article to Hansen's office mailbox. It had been written in 1981 by 3M scientists, and it described a method for measuring fluorine in blood, indicating that even back then the company was testing for fluorochemicals. One scientist mentioned in the article, Richard Newmark, still worked for 3M, in a low-lying structure nicknamed the "nerdy building." Hansen arranged to meet with him there.

Newmark, a collegial man with a compact build, told Hansen that, more than twenty years before, two academic scientists, Donald Taves and Warren Guy, had discovered a fluorochemical in human blood. They had wondered whether Scotchgard might be its source, so they approached 3M. Newmark told her that his subsequent experiments had confirmed their suspicions—the chemical was PFOS—but 3M lawyers had urged his lab not to admit it.

As Hansen wrote all this down in a notebook, she felt anger rising inside her. Why had so many colleagues doubted the soundness of her results if earlier 3M experiments had already proved the same thing? After the meeting, she hurried back to the lab to find Bacon. "He knew!" she told him.

Bacon's face remained expressionless. He told Hansen to type up her notes for him. She remembers him telling her not to email them. (In response to questions about Hansen's account, Bacon said that he didn't remember specifics. When I called Newmark, he told me that he could not remember her or anything about PFOS. "It's been a very long time, and I'm in my mid-eighties, and just do not remember stuff that well," he said.)

A few months later, in early 1999, Bacon invited Hansen to an extraordinary meeting: She would have the chance to present her findings to 3M's CEO, Livio D. DeSimone. Hansen spent several days rehearsing while driving and making dinner. On the day of the meeting, she took an elevator up to the executive suite; her stomach turned as a secretary pointed her to a conference room. Men in suits sat around a long table. Her boss, Bacon, was there. DeSimone, a portly man with white hair, sat at the head of the table.

Almost as soon as Hansen placed her first transparency on the projector, the attendees began interrogating her: Why did she do this research? Who directed her to do it? Whom did she inform of the results? The executives seemed to view her diligence as a betrayal: Her data could be damaging to the company. She remembers defending herself, mentioning Newmark's similar work in the '70s and trying, unsuccessfully, to direct the conversation back to her research. While the executives talked over her, Hansen noticed that DeSimone's eyes had closed and that his chin was resting on his dress shirt. The CEO appeared to have fallen asleep. (DeSimone died in 2017. A company spokesperson did not answer my questions about the meeting.)

After that meeting, Hansen remembers learning from Bacon that her job would be changing. She would only be allowed to do experiments that a supervisor had specifically requested, and she was to share her data with only that person. She would spend most of her time analyzing samples for studies that other employees were conducting, and she should not ask questions about what the results meant. Several members of her team were also being reassigned. Bacon explained that a different scientist at 3M would lead research into PFOS going forward. Hansen felt that she was being punished and struggled not to cry.

Even as Hansen was being sidelined, the results of her research were quietly making their way into the files of the Environmental Protection Agency. Since the '70s, federal law has required that companies tell the EPA about any evidence indicating that a company's products present "a substantial risk of injury to health or the environment." In May 1998, 3M officials notified the agency, without informing Hansen, that the company had measured PFOS in blood samples from around the U.S.—a clear reference to Hansen's work. It did not mention its animal research from the '70s, and it said that the chemical caused "no adverse effects" at the levels the company had measured in its workers. A year later, 3M sent the EPA another letter, again without telling Hansen. This time, it informed the agency about the fourteen other fluorochemicals, several of them made by 3M, that Hansen's team had detected in human blood. The company reiterated that it did not believe that its products presented a substantial risk to human health.

Hansen recalls that in the summer of 1999, at an annual picnic that her parents hosted for 3M scientists, she was grilling corn when one of the creators of Scotchgard, a gray-haired man in glasses, confronted her. He accused her of trying to tear down the work of her colleagues. Did it make her feel powerful ruining other people's careers? he asked. Hansen didn't know how to respond, and he walked away.

Several of Hansen's superiors had stopped greeting her in the hallways. When she presented a poster of her research at a 3M event, nobody asked her about it. She lost her appetite, and her pleated pants grew baggy. She started to worry that an angry coworker might confront or even harm her in the company's dark parking lot. She got into the habit of calling her husband before walking to her car.

A year after Hansen's meeting with the CEO, 3M, under pressure from the EPA, made a very costly decision: It was going to discontinue its entire portfolio of PFOS-related chemicals. In May 2000, for the first time, 3M officials revealed to the press that it had detected the chemical in blood banks. One executive claimed that the discovery was a "complete surprise." The company's medical director told *The New York Times*, "This isn't a health issue now, and it won't be a health issue." But the newspaper also quoted a professor of toxicology. "The real issue is this stuff accumulates," the professor said. "No chemical is totally innocuous, and it seems inconceivable that anything that accumulates would not eventually become toxic."

Hansen was now pregnant with twins. Although she was heartened by 3M's announcement—she saw it as evidence that her work had forced the company to act—she was also ready to leave the environmental lab, where she felt marginalized. After giving birth, she joined 3M's medical devices team. But first, she decided to have one last blood sample tested for PFOS: her own. The results showed one of the lowest readings she'd seen in human blood. Immediately, she thought of the rats that had passed the chemical on to their pups.

Hansen told me that, for the next nineteen years, she avoided the subject of fluorochemicals with the same intensity with which she had once pursued it. She focused on raising her kids and coaching a cross-country ski team; she worked a variety of jobs at 3M, none related to fluorochemicals. In 2002, when 3M announced that it would be replacing PFOS with another fluorochemical, PFBS, Hansen knew that it, too, would remain in the environment indefinitely. Still, she decided not to involve herself. She skipped over articles about the chemicals in scientific journals and newspapers, where they were starting to be linked to possible developmental, immune system, and liver problems. (In 2006, after the EPA accused 3M of violating the Toxic Substances Control Act, in part by repeatedly failing to disclose the harms of fluorochemicals promptly, the company agreed to pay a small penalty of $1.5 million, without admitting wrongdoing.)

During that time, forever chemicals gained a new scientific name—per- and polyfluoroalkyl substances, or PFAS, an acronym that is vexingly similar to the specific fluorochemical PFOS. A swath of 150 square miles around 3M's headquarters was found to be polluted with PFAS; scientists discovered PFOS and PFBS in local fish and various fluorochemicals in water that roughly 125,000 Minnesotans drank. Hansen's husband, Peter, told me that, when friends asked Hansen about PFAS, she would change the subject. Still, she repeatedly told him—and herself—that the chemicals were safe.

In the 2016 book *Secrecy at Work*, two management theorists, Jana Costas and Christopher Grey, argue that there is nothing inherently wrong or harmful about keeping secrets. Trade secrets, for example, are protected by federal and state law on the grounds that they promote innovation and contribute to the economy. The authors draw on a large body of sociological research to illustrate the many ways that information can be concealed. An organization can compartmentalize a secret by slicing it into smaller components, preventing any one person from piecing together the whole. Managers who don't want to disclose sensitive information may employ "stone-faced silence." Secret-keepers can form a kind of tribe, dependent on one another's continued discretion; in this way, even the existence of a secret can be kept secret. Such techniques become pernicious, Costas and Grey write, when a company keeps a dark secret, a secret about wrongdoing.

Certain unpredictable events—a leak, a lawsuit, a news story—can start to unspool a secret. In the case of forever chemicals, the unspooling began on a cattle farm. In 1998, a West Virginia farmer told a lawyer, Robert Bilott, that wastewater from a Du-Pont site seemed to be poisoning his cows: They had started to foam at the mouth, their teeth grew black, and more than a hundred eventually fell over and died. Bilott sued and obtained tens of thousands of internal documents, which helped push forever chemicals into the public consciousness. The documents revealed that the farm's water contained PFOA, the fluorochemical that DuPont had bought from 3M, and that both companies had long understood it to be toxic. (The lawsuit, which ended in a settlement, was dramatized in the film *Dark Waters*, starring Mark Ruffalo as Bilott.) Bilott later sued 3M over contamination in Minnesota, but the judge prohibited discussion of health repercussions; a jury ultimately decided in 3M's favor. Finally, in 2010, the Minnesota attorney general's office filed its own suit, alleging that 3M had harmed the environment and polluted drinking water. The company paid $850 million in a settlement, without an admission of fault or liability. The AG also released thousands more internal 3M records to the public.

The AG's records helped me report a series of stories for *The Intercept* about forever chemicals. Much of my reporting, which started in 2015, focused on what 3M and DuPont knew, even as they continued to produce PFAS. But, as I reported on the cover-up, I wondered what it meant for a sprawling multinational company to *know* that its products were dangerous. Who knew? How much, exactly, did they know? And how had the company kept its secret? For many years, no one inside 3M would agree to speak with me.

Then, in 2021, John Oliver did a segment on his comedy news

show, *Last Week Tonight*, about forever chemicals. The segment, which mentioned my reporting, said that they could cause cancer, immune-system issues, and other problems. "The world is basically soaked in the Devil's piss right now," Oliver said. "And not in a remotely hot way." One of Hansen's former professors sent her the segment, and Hansen watched it at her kitchen table—a moment that would eventually lead her to me.

"This actually made me sad as there are so many inaccuracies," Hansen wrote to her professor in response. But, when the professor asked her what was incorrect, Hansen didn't know what to say. For the first time, she googled the health effects of PFOS.

Hansen was deeply troubled by what she read. One paper, published in 2012 in the *Journal of the American Medical Association*, found that, in children, as PFOS levels rose so did the chance that vaccines were ineffective. Children with high levels of PFOS and other fluorochemicals were more likely to experience fevers, according to a 2016 study. Other research linked the chemicals to increased rates of infectious diseases, food allergies, and asthma in children. Dozens of scientific papers had found that, in adults, even very low levels of PFOS could interfere with hormones, fertility, liver and thyroid function, cholesterol levels, and fetal development. Even PFBS, the chemical that 3M chose as a replacement for PFOS, caused developmental and reproductive irregularities in animals, according to the Minnesota Department of Health.

Reading these studies, Hansen felt a paradoxical kind of relief: As bad as PFOS seemed to be, at least independent scientists were studying it. But she also felt enraged at the company and at herself. For years, she had repeated the company's claim that PFOS was not harmful. "I'm not proud of that," she told me. She felt "dirty" for ever collecting a 3M paycheck. When she read the documents released by the Minnesota AG, she was horrified by how much the company had known and how little it had told her. She found records of studies that she had conducted, as well as the typed notes from her meeting with Newmark.

In October 2022, after Hansen had been at 3M for twenty-six years, her job was eliminated, and she chose not to apply for a new one. Three months later, she wrote me an email, offering to speak about what she had witnessed inside the company. "If you'd be interested in talking further, please let me know," she wrote. The next day, we had the first of dozens of conversations.

When Hansen first told me about her experiences, I felt conflicted. Her work seemed to have helped force 3M to stop making a number of toxic chemicals, but I kept thinking about the twenty years in which she had kept quiet. During my first visit to Hansen's home, in February 2023, we sat in her kitchen, eating bread that her husband had just baked. She showed me pictures of her father and shared a color-coded timeline of 3M's history with forever chemicals. On a bitterly cold walk in a local park, we tried to figure out if any of her colleagues, beside Newmark, had known that PFOS was in everyone's blood. She often sprinkled her stories with such Midwesternisms as "holy buckets!"

During my second trip, this past August, I asked her why, as a scientist who was trained to ask questions, she hadn't been more skeptical of claims that PFOS was harmless. In the awkward silence that followed, I looked out the window at some hummingbirds.

Hansen's superiors had given her the same explanation that they gave journalists, she finally said—that factory workers were fine, so people with lower levels would be too. Her specialty was the detection of chemicals, not their harms. "You've got literally the medical director of 3M saying, 'We studied this, there are no effects,'" she told me. "I wasn't about to challenge that." Her income had helped to support a family of five. Perhaps, I wondered aloud, she hadn't really wanted to know whether her company was poisoning the public.

To my surprise, Hansen readily agreed. "It almost would have been too much to bear at the time," she told me. 3M had successfully compartmentalized its secret; Hansen had only seen one slice. (When I sent the company detailed questions about Hansen's account, a spokesperson responded without answering most of them or mentioning Hansen by name.)

Recently, I thought back on Taves and Guy, the academic scientists who, in the '70s, came so close to proving that 3M's chemicals were accumulating in humans. Taves is ninety-seven, but when I called him he told me that he still remembers clearly when company representatives visited his lab at the University of Rochester. "They wanted to know everything about what we were doing," he told me. But the exchange was not reciprocal. "I soon found out that they weren't going to tell me anything." 3M never confirmed to Taves or Guy, who was a postdoctoral student at the time, that its fluorochemicals were in human blood. "I'm sort of kicking myself for not having followed up on this more, but I didn't have any research money," Guy told me. He eventually became a dentist to support his wife and family. (He died this year at eighty-one.) Taves, too, left the field, to become a psychiatrist, and the trail ended there.

Last year, while reading about the thousands of PFAS-related lawsuits that 3M was facing, I was intrigued to learn that one of them, filed by cities and towns with polluted water, had produced a new set of internal 3M documents. When I requested several from the plaintiff's legal team, I saw two names that I recognized. In a document from 1991, a 3M scientist talked about using a mass spectrometer—the same tool that Hansen would use years later—to devise a technique for measuring PFOS in biological fluid. The author was Jim Johnson—and he had sent the report to his boss, Dale Bacon.

This revelation made me gasp. Johnson had been Hansen's first boss and had instigated her research into PFOS. Bacon had questioned her findings and ultimately told her to stop her work. (In a sworn deposition, Bacon said that by the '80s he had heard, during a watercooler chat with a colleague, that Taves and Guy had found PFOS in human blood.) What I couldn't understand was why Johnson would ask Hansen to investigate something that he had already studied himself—and then act surprised by the results.

Jim Johnson, who is now an eighty-one-year-old widower, lives with several dogs in a pale yellow house in North Dakota. When I first called him, he said that he had begun researching PFOS in the '70s. "I did a lot of the very original work on it," he told me. He said that when he saw the chemical's structure he understood "within twenty minutes" that it would not break down in nature. Shortly thereafter, one of his experiments revealed that PFOS was binding to proteins in the body, causing the chemical to accumulate over time. He told me that he also looked for PFOS in an informal test of blood from the general population, around the late '70s, and was not surprised when he found it there.

Johnson initially cited "480 pounds of dog" as a reason that I shouldn't visit him, but he later relented. When I arrived, on a chilly day in November, we spent a few minutes standing outside his house, watching Snozzle, Sadie, and Junkyard press their slobbery snouts against his living room window. Then we decamped to the nearest IHOP. Johnson, who was dressed in jeans and a flannel shirt, was so tall that he couldn't comfortably fit into a booth. We sat at a table and ordered two bottomless coffees.

In an experiment in the early '80s, Johnson fed a component of Scotchban to rats and found that PFOS accumulated in their livers, a result that suggested how the chemical would behave in humans. When I asked why that mattered to the company, he took a sip of coffee and said, "It meant they were screwed."

At the time, Johnson said, he didn't think PFOS caused significant health problems. Still, he told me, "it was obviously bad," because man-made compounds from household products didn't belong in the human body. He said that he argued against using fluorochemicals in toothpaste and diapers. Contractors working for 3M had shaved rabbits, he said, and smeared them with the company's fluorochemicals to see if PFOS showed up in their bodies. "They'd send me the livers and, yup, there it was," he told me. "I killed a lot of rabbits." But he considered his efforts largely futile. "These idiots were already putting it in food packaging," he said.

Johnson told me, with seeming pride, that one reason he didn't do more was that he was a "loyal soldier," committed to protecting 3M from liability. Some of his assignments had come directly from company lawyers, he added, and he couldn't discuss them with me. "I didn't even report it to my boss, or anybody," he said. "There are some things you take to your grave." At one point, he also told me that, if he were asked to testify in a PFOS-related lawsuit, he would probably be of little help. "I'm an old man, and so I think they would find that I got extremely forgetful all of a sudden," he said, and chuckled.

Out the windows of IHOP, I watched a light dusting of snow fall on the parking lot. In Johnson's telling, a tacit rule prevailed at 3M: Not all questions needed to be asked, or answered. His realization that PFOS was in the general public's blood "wasn't something anyone cared to hear," he said. He wasn't, for instance, putting his research on posters and expecting a warm reception. Over the years, he tried to convince several executives to stop making PFOS altogether, he told me, but they had good reason not to. "These people were selling fluorochemicals," he said. He retired as the second-highest-ranked scientist in his division, but he claimed that important business decisions were out of his control. "It wasn't for me to jump up and start saying, 'This is bullshit!'" he said, and he was "not really too interested in getting my butt fired." And so his portion of 3M's secret stayed in a compartment, both known and not known.

Johnson said that he eventually tired of arguing with the few colleagues with whom he could speak openly about PFOS. "It was time," he said. So he hired an outside lab to look for the chemical in the blood of 3M workers, knowing that it would also test blood bank samples for comparison—the first domino in a chain that would ultimately take the compound off the market. Oddly, he compared the head of the lab to a vending machine. "He gave me what I paid for," Johnson said. "I knew what would happen." Then Johnson tasked Hansen with something that he had long avoided: going beyond his initial experiments and meticulously documenting the chemical's ubiquity. While Hansen took the heat, he took early retirement.

Johnson described Hansen as though she were a vending machine too. "She did what she was supposed to do with the tools I left her," he said.

I pointed out that Hansen had suffered professionally and personally, and that she now feels those experiences tainted her career. "I didn't say I was a nice guy," Johnson replied, and laughed. After four hours, we were nearing the bottom of our bottomless coffees.

Johnson has strayed from evidence-based science in recent

years. He now believes, for instance, that the theory of evolution is wrong, and that Covid-19 vaccines cause "turbo-cancers." But his account of what happened at 3M closely matched Hansen's, and when I asked him about meetings and experiments described in court documents, he remembered them clearly.

When I called Hansen about my conversation with Johnson, she grew angrier than I'd ever heard her. "He knew the whole time!" she said. Then she had to get off the phone for an appointment. "So glad I'm going to see my therapist," she added, and hung up.

I once thought of secrets as discrete, explosive truths that a heroic person could suddenly reveal. In the 1983 film *Silkwood*, which is based on real events, Karen Silkwood, a worker at a plutonium plant, assembles a thick folder documenting her employer's shoddy safety practices; while driving to share them with a reporter, she dies in a mysterious one-car crash. In another adaptation of a true story, the 2015 film *Spotlight*, a source delivers a box of critical documents to the *Boston Globe*, helping the paper to publish an investigation into child sexual abuse within the Catholic Church. Talking to Hansen and Johnson, though, I saw that the truth can come out piecemeal over many years, and that the same people who keep secrets can help divulge them. Some slices of 3M's secret are only now coming to light, and others may never come out.

Between 1951 and 2000, 3M produced at least a hundred million pounds of PFOS and chemicals that degrade into PFOS. This is roughly the weight of the *Titanic*. After the late '70s, when 3M scientists established that the chemical was toxic in animals and was accumulating in humans, it produced millions of pounds per year. Scientists are still struggling to grasp all the biological consequences. They have learned, just as Johnson did decades ago, that proteins in the body bind to PFOS. It enters our cells and organs, where even tiny amounts can cause stress and interfere with basic biological functions. It contributes to diseases that take many years to develop; at the time of a diagnosis, one's PFOS level may have fallen, making it difficult to establish causation with any certainty.

The other day, I called Brad Creacey, who became an Air Force firefighter in the '70s at the age of eighteen. He told me that several times a year, for practice, he and his comrades put on rubber boots and heavy silver uniforms that looked like spacesuits. Then a "torch man," holding a stick tipped with a burning rag, ignited jet fuel that had been poured into an open-air pit. To extinguish the hundred-foot-tall flames, Creacey and his colleagues sprayed them with aqueous film-forming foam, or AFFF. 3M manufactured it from several forever chemicals, including PFOS.

Creacey remembers that AFFF felt slick and sudsy, almost like soap, and dried out the skin on his hands until it cracked. To celebrate his last day on a military base in Germany, his friends dumped a ceremonial bucket on him. Only later, after working with firefighting foam at an airport in Monterey, California, did he start to wonder if a string of ailments—cysts on his liver, a nodule near his thyroid—were connected to the foam. He had high cholesterol, which diet and exercise were unable to change. Then he was diagnosed with thyroid cancer. "It makes me feel like I was a lab rat, like we were all disposable," Creacey told me. "I've lost faith in human beings."

It may be tempting to think of Creacey and his peers as unwitting research subjects; indeed, recent studies show that PFOS is associated with an increased risk of thyroid cancer and, in Air Force servicemen, an elevated risk of testicular cancer. But it is probably more accurate to say that we are all part of the experiment. Average levels of PFOS are falling, but nearly all people have at least one forever chemical in their blood, according to the Centers for Disease Control and Prevention. "When you have a contaminated site, you can clean it up," Elsie Sunderland, an environmental chemist at Harvard University, told me. "When you ubiquitously introduce a toxicant at a global scale, so that it's detectable in everyone . . . we're reducing public health on an incredibly large scale." Once everyone's blood is contaminated, there is no control group with which to compare, making it difficult to establish responsibility.

New health effects continue to be discovered. Researchers have found that exposure to PFAS during pregnancy can lead to developmental delays in children. Numerous recent studies have linked the chemicals to diabetes and obesity. This year, a study discovered thirteen forever chemicals, including PFOS, in weeks-old fetuses from terminated pregnancies and linked the chemicals to biomarkers associated with liver problems. A team of New York University researchers estimated in 2018 that the costs of just two forever chemicals, PFOA and PFOS—in terms of disease burden, disability, and health-care expenses—amounted to as much as $62 billion in a single year. This exceeds the current market value of 3M.

Philippe Grandjean, a physician who helped discover that PFAS harm the immune system, believes that anyone exposed to these chemicals—essentially everyone—may have an elevated risk of cancer. Our immune systems often find and kill abnormal cells before they turn into tumors. "PFAS interfere with the immune system, and likely also this critical function," he told me. Grandjean, who served as an expert witness in the Minnesota AG's case, has studied many environmental contaminants, including mercury. The impact of PFAS was so much more extreme, he said, that one of his colleagues initially thought it was the result of nuclear radiation.

In April, the EPA took two historic steps to reduce exposure to PFAS. It said that PFOS and PFOA are "likely to cause cancer" and that no level of either chemical is considered safe; it deemed them hazardous substances under the Superfund law, increasing the government's power to force polluters to clean them up. The agency also set limits for six PFAS in drinking water. In a few years, when the EPA begins enforcing the new regulations, local utilities will be required to test their water and remove any amount of PFOS or PFOA which exceeds four parts per trillion—the equivalent of one drop dissolved in several Olympic swimming pools. 3M has produced enough PFOS and chemicals that degrade into PFOS to exceed this level in all of the freshwater on Earth. Meanwhile, many other PFAS continue to be used, and companies are still developing new ones. Thousands of the compounds have been produced; the Department of Defense still depends on many for use in explosives, semiconductors, cleaning fluids, and batteries. PFAS can be found in nonstick cookware, guitar strings, dental floss, makeup, hand sanitizer, brake fluid, ski wax, fishing lines, and countless other products.

In a statement, a 3M spokesperson told me that the company "is proactively managing PFAS," and that 3M's approach to the chemicals has evolved along with "the science and technology of PFAS, societal and regulatory expectations, and our expectations of ourselves." He directed me to a fact sheet about their continued importance in society. "These substances are critical to multiple industries—including the cars we drive, planes we fly, computers and smart phones we use to stay connected, and more," the fact sheet read.

Recently, 3M settled the lawsuit filed by cities and towns with polluted water. It will pay up to $12.5 billion to cover the costs of filtering out PFAS, depending on how many water systems need the chemicals removed. The settlement, however, doesn't approach the scale of the problem. At least 45 percent of U.S. tap water is estimated to contain one or more forever chemicals, and one drinking water expert told me that the cost of removing them all would likely reach $100 billion.

In 2022, 3M said that it would stop making PFAS and would "work to discontinue the use of PFAS across its product portfolio," by the end of 2025—a pledge that it called "another example of how we are positioning 3M for continued sustainable growth." But it acknowledged that more than sixteen thousand of its products still contained PFAS. Direct sales of the chemicals were generating $1.3 billion annually. 3M's regulatory filings also allow for the possibility that a full phaseout won't happen—for example, if 3M fails to find substitutes. "We are continuing to make progress on our announcement to exit PFAS manufacturing," 3M's spokesperson told me. The company and its scientists have not admitted wrongdoing or faced criminal liability for producing forever chemicals or for concealing their harms.

Hansen often wonders what her father would say about 3M if he were still alive. A few years ago, he began to show signs of dementia, which worsened during the Covid-19 pandemic. Every time Hansen explained to him that a novel coronavirus was sickening people around the world, he asked how he might contribute—

forgetting that the N95 mask he helped to create was already protecting millions of people from infection. When he died, in January 2021, Hansen noticed some Coban wrap on his arm. It was shielding his delicate skin from tears, just as he had designed it to. "He invented that," Hansen told the hospice nurse, who smiled politely.

After she left 3M, Hansen began volunteering at a local nature preserve, where she works to clear paths and protect native plants. Last August, she took me there, and we walked to a creek where she often spends time. The water is home to three species of trout, she told me. It is also polluted by forever chemicals that 3M once dumped upstream.

For most of our hike, a thick wall of flowers—purple joe-pye weed and goldenrod—made it impossible to see the creek bank. Then we came to a wooden bench. I climbed on top of it and looked down on the creek. As I listened to the gurgling of water and the buzzing of insects, I thought I understood why Hansen liked to come here. It was too late to save the creek from pollution; 3M's chemicals could be there for thousands of years to come. Hansen just wanted to appreciate what was left and to leave the place a little better than she'd found it.

MAX G. LEVY

I Tried to Train My Color Vision. Here's What Happened.

FROM *Sequencer*

ONE AFTERNOON DURING my PhD, I took a break from lab work to snack on a banana. I grabbed a seat in the office, slid off my headphones, and peeled open my treat. Just as I bit in, I noticed a lab mate staring at me.

"Max," she told me, holding back some laughter. "That banana is a four."

I instantly knew what she meant because I'd made this mistake before. The banana was days from being ripe, and I had misjudged the color. A laughably green banana.

Even before I knew that I had mild deuteranomaly (so-called red-green colorblindness), I struggled with cryptic color schemes on spreadsheets and graphs. Whether in spite or because of this, color theory fascinated me. I had neurological, practical, and philosophical questions. Why do we call the retina's longest wavelength cone "red" when it actually best absorbs yellow-green light? Why does mixing paint obey different rules than mixing light? If I could see through your eyes, would your mental images match mine—does your blue match my blue? And I had questions that blended all three, like *what the hell is brown???*‡

Imagine my delight, then, when I found a Wordle-style color-guessing game called Hexcodle.

The name is a portmanteau of Wordle and "hexadecimal code," the sixteen-digit language used to program digital colors. Your phone and TV display millions of colors by mixing precise intensities of red, green, and blue light. The code represents each intensity with a range of sixteen digits—0,1,2,3,4,5,6,7,8, 9,A,B,C,D,E,F. Each color usually runs from 0 to 255—00 to FF. #FF0000 codes for red, #00FF00 for green, and #0000FF for blue. And mixing all the lights (#FFFFFF) or none of the lights (#000000) in RGB gives white and black, respectively.

Hexcodle is meant to appeal to graphic design nerds. Yet as someone who struggles to tease out subtle elements in color—a dash of red in a sea of blue, or a sprinkle of green in a stone gray—I clung to Hexcodle. In doing so, I posed a provocative question: Could I train myself to improve my color vision?

I wanted to excel in this game despite my limits. And I wondered whether I could combat those same limits. Over weeks of playing it, the simple game challenged my color vision in ways I never expected. And it made me rethink what I—and everyone else—are seeing when we look at colors.

Plastic Eyes

We don't understand color vision as well as you'd think from decades of academic study. We have classic theories of how red, green, blue, yellow, bright, and dark encode in the brain, Mike Webster of the University of Nevada, Reno, told me. "That's still the theory you get in textbooks. But it's a very naive theory," he said. "There's a whole mystery of how the brain really represents color."

I began my quest by phoning the color blindness expert Jay Neitz. Neitz has led studies with his wife, Maureen, for more than thirty years. In a pivotal study from 2009, their team at the University of Washington cured color blindness in monkeys—a feat so surprising that it earned them *Time* magazine's third-best scientific discovery that year.

Neitz's *Nature* study exploded minds for a few reasons. Monkeys (and people) usually have three cones that absorb short (blue/violet), medium (green), and long (yellow-green) wavelengths. Most color deficiencies come from anomalies in the cones, but one form called "dichromatism" is a genetic condition where one cone is entirely missing. The Neitzes treated dichromat monkeys with this categorically severe version of color blindness. And not only did they replace the missing type of cone with a first-of-its-kind gene therapy, they did this in *adult* monkeys, raising entirely new questions about sensory plasticity.

"Most people believed that you'd have to do it in a very young monkey to get them to be able to take advantage of the new cone," Neitz said.

Now . . . let me just speak for myself, I'm not getting this treatment anytime soon. The therapy is far from approval and, more importantly, I'm not so deficient that I want an eye injection that temporarily detaches the retina. (Neitz's lab currently investigates more palatable delivery methods.)

What instead compelled me about this work was the sort of neuroplasticity here. Neuroplasticity is basically the idea that the brain forges new connections when existing ones leave it deficient. It's an innate flexibility that runs wild in your developmental years and wanes with age. Yet something was happening in these adult monkeys' brains that meant they could go from lacking a cone for "normal" vision to quickly accepting a new anatomical input.

On one hand, this discovery revealed a stunning degree of plasticity. But on the other, I had no plans to change my retinas—I wanted to squeeze more out of the parts I've got—and this study highlighted just how important it is for color vision to have all the parts in optimal working order. *Gulp.*

So I asked Neitz more directly whether he thinks color vision is plastic enough to train. He hesitated for several seconds, then affirmed that our visual systems constantly adjust to what our eyes experience.

"We learn through our whole entire lives. That's also a kind of plasticity," Neitz said.

Neitz demonstrated two decades ago that study participants who wore red-tinted goggles for a few hours every day became less sensitive to red light. Here's an excerpt from the results, bold emphasis is my own.

"*. . . each subject's unique yellow shifted progressively further away from his or her baseline. Initially, the size of the shift was small; however, after many days of exposure, for example, to the red alteration conditions, the shift had grown so large that* **wavelengths previously called red came to be consistently called green** *in appearance.*"

These were adult humans, and in another stunning display of plasticity, it took about a week for participants' color vision to return to normal. But this evidence of lasting changes with repetition fanned the embers of my optimism.

One route for improving my color vision, albeit temporarily, could be an over-the-counter remedy. Commercial colorblind glasses block out confusing colors between (often long-medium) cones that overlap more than normal, but there's limited evidence of long-lasting perceptual shifts caused by frequent use.

It's by design that our visual systems recalibrate incessantly, correcting for changes in the world and changes in ourselves. The same neurological "settings" that let you see on a bright day don't work when you step into a matinee movie. Even as eye lenses yellow with age, the visual system compensates to produce a clear picture. Our perception of colors and ability to discriminate between them changes remarkably little with age, according to John Barbur, a professor of optics and visual science with City St George's, University of London. "Color vision is actually arguably the most robust attribute," he said.

"As you age, all kinds of horrible things are happening to your visual system—it's falling apart," said Webster, who wrote a scientific review about compensation in color vision. "You don't want to see the world falling apart."

So while color discrimination varies from person to person, including among people without deficiencies, the one constant is an individual brain's desire to extract as much information as possible from a finite number of cells, and keep that information consistent.

That's where neurological adjustments and compensation shine. Think of it like tweaking contrast on a photograph or focusing a camera lens. Proper adjustments reveal more edges and details—more useful information. Similarly, when color differences are subtle, our adjustments can help discern finer differences, correcting for a yellowing filter or muted colors, for example. The adjustments are volume knobs on a neuron's activity levels.

We know that something like this happens because people with anomalous trichromacy can discriminate colors better than their cones may suggest, and not just because they've learned which colors "should be" most different. When Webster's team displayed colors to people in brain scanners in a 2021 study, they reported stronger associated brain activity in people with deficient cones compared to people with functional ones. "It really is a sensory magnification and not just a learned way of reporting the world," Webster said.

There's no evidence that we can train away color deficiencies but, theoretically, practice can improve vision perception in general. "There's no question, you can train yourself to become better." Webster said. "By forcing yourself to make these fine changes and getting the feedback on what's correct, you can really learn to see better."

Hearing that excited me, but I couldn't help wondering whether I could notice any gains.

Gaming My Vision

Within my first few days playing Hexcodle, I noticed my color deficiency sneaking up on me.

Some of my supposedly incorrect guesses looked indistinguishable (to me) from the correct target. On Hexcodle #438, for example, I saw a bright blue just slightly off from pure. I guessed "0A" for red, "39" for green, and "E3" for blue. (Two hexadecimal digits cover 0 to 255. So this RGB guess translates to "10," "57," and "227" in conventional base-10 numbers.)

My guess seemed quite close. But it was way off. So I guessed again based on the feedback: #3535FA. Still wrong but now I *really* couldn't distinguish my guess from the target. It took me two more guesses to find the right answer: #5336F7, or "Meteor Shower." Put differently, my first guess of red intensity was about 4 percent, but the correct red intensity was 33 percent.

It felt like the only way to maneuver toward a correct answer was to use hints or guess wildly. How could I expect to learn the lesson when I couldn't *see* the lesson?

I should caveat for a sec. There are 16,777,216 unique six-digit hex codes. If you feel it's unrealistic to pick the exact code on the first or second try . . . join the club: That'd be a bonkers color acuity. Borderline sorcery. I agree.

Unfortunately, you and I are wrong. Some people can do it. Hexcodle developers Ekim Karabey and Hannah Larsen added user-requested "Hard" and "Expert" modes that limit and eliminate hints. And TikToker Jared Cross has gone viral for his skills nailing colors immediately. In one video, he appears to play the game with inverted colors.

I assumed that as the eyes behind Hexcodle, Karabey and Larsen would excel at the game. But Larsen confessed that's a common misconception. "We're just as clueless as everyone else," she said.

Larsen and Karabey both prefer the "Mini" version of the game, which only requests one digit for each red, green, and blue channel. "I just feel like my eyes aren't well trained enough. I'm literally not able to tell the difference to that granular extent," Karabey said.

In more neurological words, what he might mean is that his brain can't amplify a signal that it can't detect in the first place.

But how much do we really know about how our brains process color? Increasingly, the inner workings of color vision are being questioned.

Scientists' classic model for human color vision is opponent process theory, which proposes that we build color information from different scales: light vs. dark, blue vs. yellow, and red vs. green. If you imagine each quality existing on an axis, you wind up with a multidimensional space of limitless color combinations. It's intuitive from retinal anatomy because these colors correspond roughly to our cones. But it also implies that pure red and green, for example, are singular characters inside the brain. That's where the theory starts to fray.

Scientists have yet to find cells in the brain that directly represent the red-green and blue-yellow comparisons predicted by the theory. "Yellow, red, green, and blue should be special," Webster said. "But the more we look for them, we can't find a place in the brain where those colors are any more represented or in a different way than other hues like orange or purple."

Even the standard roles of your hordes of cones, rods, and photosensitive ganglion (which contribute to circadian rhythms) get muddled in real vision. These different "channels" of vision don't perform in isolation. When one fails, the other signals become more important. It's a sort of compensation, or adaptation, or plasticity. "To me, it's just making best use out of what there is," Barbur said.

By extension, there's no way to drill one of these players without engaging the whole system.

My takeaway was that you can train day and night on color vision tests—really devote attention toward piecing together camouflaged numbers—and you may improve at that task. But the system is so complex that we don't yet know what is improving *biologically*.

And, for now, we can't say whether training in color tasks can improve color vision in general. In Webster's 2020 review, he and his co-authors propose that the limits of plasticity are still equally mysterious: "Understanding why compensation is incomplete is as important as understanding why it occurs at all," they wrote.

My Hexcodle performance has improved slightly, but I can't say whether that means I have better color acuity. I want to believe I'm more perceptive. But is it because I've improved some aspect of my biology, or am I just more attentive? When I stare at a color, a conveyor belt of questions rolls into my psyche: Is there green in this blue? Red? Is it pale? I ask myself whether gray is an absence of color.

Just as the yellow you see on your phone is actually a sneaky combo of red and green pixels, gray is what happens when red, green, and blue intensities match. So colored light comes from imbalance. Migrating from #777777 ("Lucky Grey") to #7777BB presents a blue-tinged "Stormy Horizon." Move down to #777733, and you'll find "Garden Weed."

This renewed perspective helps me imagine what happens in my cones, deficiencies and all. (Bless their little hearts.) It may help as I continue seeking gains. "The only way you know what the rules are about—what plasticity there is—is to do the experiment," Neitz told me.

Neitz's latest research explores what "extra dimension" may arise when people (mainly women) have a fourth cone. The benefit of tetrachromacy remains a mystery since trichromatic vision has guided so much of our world, from paints to LEDs to language.

Though we lack the vocabulary to describe that extra dimension, tetrachromats may find it the boundless spectra of nature: floral pigments, soils, and of course underripe bananas.

Maybe tetrachromats can see something in a banana peel that you don't. Then again, maybe one day I will too.

‡ *Brief simplified answers to the above questions FYI*
1. Scientists today refer to cones as short-, medium-, and long-wavelength. Each absorbs the most light around 420 nm, 530 nm, and 565 nm—violet, green, and yellow-green, respectively. But that's just the peak absorption. They also absorb light away from that peak, and the L cone absorbs red light well.
2. My easy rule of thumb is that pigments and paint are "subtractive," because we only see colors that are reflected and not absorbed. So if you mix blue paint that absorbs everything except blue with red paint, you're still absorbing the other colors (**ROYGBIV**), you're getting darker because you reflect relatively less blue and red (red paint absorbs blue wavelengths and vice versa), and the residual reflected blue and red light combine for something purplish. When you mix light, red and blue still make purplish hues, but combining them adds brightness. (Pure red + pure green + pure blue = white light (#FFFFFF).)
3. Idk dude!!
4. Basically dark orange

EMMA MARRIS

New York Is Wilder Than You Think

FROM *The New York Times*

NEW YORK IS about to get really wild. Every spring, millions of birds fly from South and Central America along the Atlantic Flyway, heading for northern breeding grounds. Many of them choose to fly through New York City, sometimes stopping to rest in Central Park. You can take the subway to 81st Street, stroll across the way, and maybe see a maraschino-cherry-red male scarlet tanager with black wings, fresh from the foothills of the Andes.

As the migration picks up in the weeks to come, millions of wild birds will pass through the city. But New York is also home to wildlife year-round. Humans share the city with hundreds of species of wild animals, from red-tailed hawks and coyotes to pigeons and rats.

It might sound odd to call a pigeon or a rat a wild animal. These notoriously urban animals might seem too much creatures of the human world. If they take the subway and eat churros and pizza, can they truly be called wild? But red-tailed hawks and coyotes eat these dubiously wild rats and pigeons—so does that mean they aren't wild either?

The confusion arises because "wild" can be defined in several sometimes contradictory ways. One definition is "uninfluenced by humans." A wild place is a place that humans have not shaped. A wild animal is an animal whose life is led outside the sphere of human influence. But if we define wildness as "uninfluenced by humans," then there is probably no complete wildness left. Thanks to climate change, pollution, and the vast conversion of land for food production, all animals and all places are now influenced by humanity, at least to some degree.

But this definition of "wild" is premised on the idea that humans are categorically different from other species, that we are somehow outside nature, despite being very clearly and closely related to other animals. We may be very fancy apes, with our iPhones and our airplanes, but apes we remain. Why does just one kind of ape have this magic power of reducing or destroying wildness?

In addition to being based on an unscientific division of humans from all other animals, the idea of wildness is tangled up with a long history of racism. In the Americas, land that was managed by Indigenous people was often characterized by colonists as "wild," either out of ignorance of how management like prescribed burning, managed hunting, planting, and tending had shaped the landscape or out of an unwillingness to see Native people as having any agency at all. Characterizing people's homes as wilderness implies that the people who lived there weren't human.

There is an alternative. We can define "wild" in terms of the level of autonomy of the individual we are describing. If an animal is free to make its own choices—even in a world shaped by humanity—then it is wild. If its life is controlled by another organism, then it is not wild. Under this definition, pets, farm animals, and zoo animals are not wild. Neither is an aphid that is farmed for honeydew by a colony of ants. Humanness is not central to this definition.

I use the breakfast test. Does the animal choose its own breakfast? Then it is wild. It doesn't matter if the animal is "native" to a place. It doesn't matter if the place itself has been radically transformed, like New York City. Indeed, a single animal can change its status. Flaco, the Eurasian eagle-owl who was freed from captivity at the Central Park Zoo and lived free for a year before his recent death, was not wild for thirteen years. When he was let out and began making his own decisions—choosing between rat and pigeon for breakfast instead of just eating whatever was provided by the zoo—he became wild. Wildness is freedom.

Under this definition, New York and cities everywhere are bursting with wildness. The birds migrating through Central Park are wild. The rats raising generations of young in the park are also wild. We are surrounded by other creatures with their own agendas going about their sometimes-mysterious business. For me, this simple revelation changes the texture of the city. I see it now as an ecosystem of wild creatures that just happens to also feature lots of concrete and glass.

I believe the wildness of individual autonomy is valuable because it is a component of animal flourishing. A poodle may be happier in captivity, but many animals want to be free. You didn't see Flaco desperately trying to return to his cage.

Many people value wildness. But which kind of wildness do they mean? Do they mean that they value things that are not influenced by humans? Do they just hate anything that humans have touched? I don't think love of the wild is usually so simple-mindedly misanthropic. I think for most people, loving the wild means respecting the well-being and independence of other species and seeking to be humble, to step back from being in control. Defining "wildness" as individual autonomy captures those ethical features.

If we care about individual autonomy rather than the overall level of human influence, we should not micromanage our "wildlife" in an attempt to replicate some kind of prehuman ecosystem. Nor should we police where animals live in the name of "native species" purity, although we might sometimes decide to manage them for human safety or for other goals, such as saving vulnerable species they might like to eat.

Living together is always a compromise, and no autonomy is complete. In human societies we agree to limit our own freedoms to better coexist, and we should make compromises in interspecies relationships too. We can turn down the lights during migration seasons to prevent millions of wild birds from becoming disoriented to the point of fatal exhaustion or from hitting buildings, even if it is slightly inconvenient for us humans. On the other hand, we have a right to control insects, rats, and mice where we eat and store food for our own health and safety. But where we do decide to infringe on other animals' autonomy, we should do so as a last resort, and within a context of overall respect—and, ideally, reciprocity.

Respecting wildness is good for our character. We shouldn't be managing every other species. For one thing, we aren't smart or knowledgeable enough to do it perfectly. There's a lot about ecology we still don't understand. We should be humble and let other organisms make their own choices, except in situations where we choose to intervene to protect ourselves or other species.

Someone recently asked me whether humans can be wild, under my definition. That's a tough one, but I think we can, at least sometimes. The last time I was in New York City, it was fall. I visited Flaco, napping the afternoon away in a tree in Central Park. I found a pile of rat bones under his perch. A wild life is not always a safe life. Afterward, I wandered south with no destination in mind. I looked at the zippers and buttons in the windows in the garment district. I walked by the Strand bookstore and bought a used novel from the sidewalk racks just because I liked the title. I sat in Washington Square Park and eavesdropped on two college students talking about their futures. For an afternoon, at least, I was as wild as a migrating cerulean warbler, as wild as a New York pigeon.

JASON MAST

This Father Built a Gene Therapy for His Son. Now Comes the Harder Part: Saving Others' Children, Too.

FROM *STAT*

TERRY PIROVOLAKIS RECEIVES roughly five to ten messages a week from strangers. Sometimes, his inbox fills overnight with emails from China. Sometimes, he'll end a string of Zoom meetings to find texts from across the U.S.

Their stories are as variable as our 3-billion-letter genomes and the still-uncounted ways they can go wrong: muscles crumbling, neurons misfiring, eyes fading to black. Yet they all have the same request: My child faces a devastating rare disease, can you help us?

One of those emails came late at night last year from the sleepy Denver suburb of Littleton, where Rebekah Lockard lay curled around a beanbag. Lockard's one-year-old daughter, Naomi, had just been diagnosed with hereditary spastic paraplegia-50, or SPG50, a condition known to affect around 100 people on the planet.

The news fractured the family's life. One moment, Lockard, eight months pregnant, was browsing matching green pajamas for the soon-to-be family of four. Then her phone pinged with Naomi's genetic test results and she was drowning in scientific jargon. She understood only a few words: "paraplegia," "epilepsy," "severe intellectual disability." An hour later, doctors told her there was no cure and a 25 percent chance their unborn child had it too. After that, it didn't take much googling to find Pirovolakis.

Pirovolakis isn't a scientist or a doctor. He doesn't have vast wealth to give out—he has flirted with financial ruin. He's a forty-five-year-old IT professional, with three children, a fondness for monochrome T-shirts, and a modest house in Toronto. Yet he has become, improbably, the first stop and last hope for many families who have watched for-profit drugmakers abandon the conditions that affect their children.

His elevation is a symptom of just how much the traditional drug development system for rare diseases has broken down. In the last two decades, researchers have made huge strides in their ability to diagnose one-in-a-million mutations and devise treatments that patch or replace garbled DNA, with gene therapy, CRISPR, or RNA-based medicines. Dazzled by the new technology, investors and pharma giants initially poured billions into the field.

Then they got a hard look at the business case. While a potential therapy can now be whiteboarded in hours, the actual engineering, manufacturing, and testing still takes years and tens of millions of dollars, with no guarantee it will work. The vast majority of these conditions affect fewer than 350 patients in the U.S., hardly a blockbuster market. Even companies that charged $3 million a dose didn't necessarily make a profit. Over the last few years, companies have shelved or deprioritized over fifty gene therapies, leaving the families who had been counting on them adrift.

In some cases, companies encountered significant safety or efficacy hurdles—these treatments, while often powerful, are rarely panaceas, especially when given to older children or adults. But in most cases, it came down to a basic financial calculus. Sometimes, seemingly curative treatments were abandoned.

Pirovolakis is perhaps the most successful of a new wave of parents trying to take the tools of genetic medicine into their own hands and will treatments into existence. At first his goal was mainly to save his one-year-old son Michael, who was diagnosed in 2019 with SPG50, the same disease the Lockards would face. The plan was to build a treatment for Michael and then hand it to a company with the capital to bring it to market.

He was a stunningly successful fundraiser. He worked with university scientists in Texas, toxicologists in Quebec, and drug manufacturers in Spain to get Michael treated in under three years, a possible record. But by then, no company was interested. Pirovolakis watched children struggling, while gene therapies for their rare conditions stagnated in academia or were left sitting in industry freezers.

He became an informal counselor to other parents who, seeing no alternative, tried to follow his lead. He looked at their children's genetic test results and told them what might be possible. He helped them negotiate research agreements with university scientists or navigate the arcane details of drug manufacturing.

And he decided he would have to go further: He founded his own biotech and vowed to take to market Michael's treatment and at least four other gene therapies too rare for industry. He envisioned one day doing dozens. There are, after all, thousands of genetic disorders. Each are vanishingly uncommon but collectively they affect millions.

The business model he devised was, he admits, "crazy." He had to rely on GoFundMe donations and the fickle largesse of philanthropists and government bureaucrats. And he had to live with impossible choices, like the one presented by the Lockards.

That night, in May 2023, he responded to Rebekah in minutes. They spoke the next morning. Pirovolakis promised to do everything in his power to get Naomi treated and, if it came to that, her next child too.

Yet when Rebekah's son, Jack, was born and diagnosed, Pirovolakis only had enough resources to treat one baby. The other would have to wait.

A Victory for Michael

Before Pirovolakis found gene therapy, he flew with his wife, Georgia, and a nine-month-old Michael to Greece, where Pirovolakis's family is from. They knelt in front of the old tomb of the healer St. Nektarios on the island of Aegina, Terry cradling Michael in his arms, and prayed the doctors would ultimately find nothing.

Michael was their third child. After an uneventful birth, they were home in three hours. But he was unusually calm. And after five months, he still couldn't lift his hands. Doctors checked for Zika virus and other semi-common causes, but found nothing. Finally, after an MRI found brain malformations, Canada approved full genome sequencing.

Pirovolakis barely remembered the meeting at Toronto SickKids, only the upshot: His son would suffer from a devastating, incurable disease. After they got home, he went for a walk and collapsed in tears on the side of the road.

By the next morning, he started forming a plan.

"I wanted to eradicate the disease," he said. "You want to take it out on something. Unfortunately, it's not a person or a country, it's an invisible disease. So what I could do is eradicate it, so it never affects another child."

Searching online, he found the website of a family raising money to develop a genetic medicine for a related disease, SPG47. A few dozen families were already trying to bootstrap therapies. They were inspired by Lori Sames and Steve Gray, a North Carolina mom and academic who built a gene therapy for an ultra-rare neuron disease that affected Sames's daughter, Hannah. Most never raised enough money.

Pirovolakis, though, would prove incredibly effective. Six foot three and burly, with tufts of wayward brown-gray hair, he had earnest eyes and was disarmingly soft-spoken. People—scientists, donors—wanted to help him. When they didn't, he had a Barnum instinct for grabbing attention. He was also willing to risk every dollar he had.

He sat down with Georgia and told her that if they did this, he would have to make it his life. They might go broke. He might be essentially unavailable as a father for years.

"She said, do whatever you need to do," he said. His boss at a Canadian bank said much the same.

Within four days, he refinanced the house, liquidated his assets, and offered the CEO of SickKids $4.5 million he didn't have. When the CEO declined, saying they didn't have the capacity, Pirovolakis flew to Boston. He showed up at the American Society for Cell and Gene Therapy conference, armed with a poster that read, in the Old West style: "WANTED: a cure for Michael. REWARD: Research Grant." It had a photo of Michael smiling in ambulance-and-police-car pajamas. Pirovolakis posted it around the conference, while scanning for a face he'd memorized online.

"He more or less ambushed me," said Steve Gray.

Gray, who now works at UT Southwestern Medical Center, had already turned Pirovolakis away by phone. After the Sameses' story, he acquired "something of a reputation," said Gray. People emailed him weekly, detailing diseases and genes he had never heard of. He couldn't say yes to everyone.

He also had a hard time saying no. Cornered, he pulled out a sheet showing each step and its cost and warned him there were no guarantees. Okay, Pirovolakis said.

The next three years were a blur. Journalists paraded through the house. He biked to Ottawa to meet the prime minister, held galas and golf tournaments, sold donuts to drunks outside a concert. Meanwhile, Michael said his first words and almost started walking. Then he suffered a three-hour seizure. Doctors weren't sure he'd wake up. When he did, he couldn't speak.

In late 2021, they had an appointment to sell their house when an adviser, a biotech executive, called. The executive tried, unsuccessfully, to get his company to take on the therapy, but it was just too rare. So he sold $1 million of stock. Take it, he told Pirovolakis.

They dosed Michael on a warm day in March. Pirovolakis sat with Georgia outside the OR, full of terror, uncertain if he had helped or hurt. But when Michael awoke two hours later, healthy and unbothered, Pirovolakis sighed with relief. He figured he was done.

"I thought, I'll get it to the clinic, I'll show that Michael can be treated, and then someone must want this," he said. "You know, someone out there is going to take this on. And then, no one did."

Tests for Naomi and Jack

At first, Rebekah and Evan thought they won the parental lottery. Naomi was so serene. You could take her out to dinner, set her in a bassinet, and she would lie untroubled, a small round face blinking with huge blue eyes.

The couple met at CU Boulder. Rebekah was in law school, on the path to becoming a child advocate. Evan was an engineering student making his second crack at college after two lonely years manning his family's West Virginia farmland. They dated on the slopes, Evan snowboarding while Rebekah skied.

After Naomi was born, Evan started ski lessons, so they could teach the children together. Rebekah imagined showing them Europe, a privilege she never had.

Then Naomi remained serene. The silence grew eerie the more time passed. She barely cried and didn't crawl or roll. Rebekah knew something was off, though about a year would pass before doctors suggested a full genome test.

Jack was born four weeks after Naomi's diagnosis, a month of sheer, joyless terror. At first, they were elated.

"I was convinced he didn't have it," said Evan. "He was just kicking and screaming."

It would be weeks before genetic test results came back. In the meantime, Jack struggled to feed. He became severely underweight. One night, he aspirated on formula. They took him to the hospital, where, in the mayhem of a pediatric ward, as they lay on a couch watching Jack breathe through an oxygen tube, Rebekah's phone buzzed: test results.

"I couldn't even grasp it," she said. She turned to Evan. "I don't know, I can't figure out if he has it or not."

"Well," Evan said. "It says 'positive.'"

Rebekah texted Pirovolakis, who called instantly. "We're going to do it," he told her. "We're going to do it.'"

Terry's Gambit

By the time Jack was diagnosed, in July 2023, Pirovolakis had long since accepted that no one was coming to license his treatment. Companies were abandoning gene therapies. Pfizer shut down its division. Amicus Therapeutics nixed a $400 million spinout that would've developed eight treatments, with rights to dozens more. One company, called Taysha, licensed fifteen gene therapies from UT Southwestern, many of which Steve Gray had been building with help from families, went public, reached a $1 billion valuation, and then dropped all but one. Parents were left fighting to regain rights to the treatments.

The withdrawal was so sudden that, in certain facilities, clinical-grade gene therapies were left in deep freeze, all but ready to be put into patients if only someone would pay for the trial—like a fully stocked grocery store hastily abandoned in some post-apocalyptic film.

One of Pirovolakis's first solutions was to call up old supporters and ask for $5 million to $10 million to develop it himself. They blanched. Donors, Pirovolakis learned, might write $100,000 checks. But there's a reason for-profit companies develop virtually all drugs.

So, in September 2022, Pirovolakis founded his own company. Elpida Therapeutics, he called it, after the Greek word for "hope." He laid out the vision to Javier Garcia, a longtime pharma executive and philanthropist, at a soccer stadium, where a tournament was being held in Michael's honor. "I have a crazy idea," he began.

Elpida would develop several gene therapies, he explained. Most of the conditions would be too rare to profit. But there was an obscure government voucher program—one industry often ignored, because it was uncertain, subject to repeal, and "only" worth around $100 million—for companies that got drugs approved for rare pediatric diseases.

If he got one approved, he could use the voucher to pay back investors and then funnel the rest into his other programs. Garcia gave him a loan to hire consultants and a four-person team.

At the same time, the NIH and FDA were setting up their own efforts. They wanted to make gene therapy more of a platform. In theory, the technology was modular. Researchers used small, nonpathogenic viruses to shuttle genes into patients' cells. Different viruses travel better to different tissues. For a given disease, just pick the best virus for the affected organs and slot in the given gene.

The reality was far messier and costlier, for reasons encompassing the gritty nuances of each individual disease and expensive animal experiments and manufacturing. Through the Bespoke Gene Therapy Consortium, or BGTC, government scientists planned to push gene therapies for several ultra-rare diseases into trials, and in the process, write a "playbook" for future advocates, academics, or companies to follow.

Pirovolakis became a key member. At conferences, you could see the former IT professional huddled with the NIH elite. But the BGTC was only designed to get treatments into trials. No one knew how to marshal the resources to get them approved.

Elpida was one uncertain attempt. "If this model works, it could really be a roadmap," said Amber Freed, who is trying to raise money to develop a therapy Taysha dropped, for her son's rare form of epilepsy.

Others were less sanguine. Matt Wilsey, a tech entrepreneur who founded a company around his daughter's ultra-rare disorder, questioned how scalable it was. For Michael's treatment, Pirovolakis's gregarious charm earned him discounts and pro bono work for manufacturing and animal studies. Could you scale that?

"I don't know how realistic that is," Wilsey said.

Neither Pirovolakis nor his collaborators disagreed. "It's not a good model," said Gray. "It's just that what we have available doesn't work."

After Michael was treated, there was still virus left over. With the cash from another SPG50 nonprofit, Pirovolakis treated a child in Spain and a child in Connecticut. He planned to then start a clinical trial for eight more patients, to gather data for approval.

Then Jack Lockard was diagnosed. For Michael and the other children, gene therapy wouldn't be close to a cure. Too much damage had been done and the virus doesn't reach every neuron. But an infant might just be young enough. And there was just enough virus left.

Somewhere in Between

On a sunny day last spring, Evan lay on the beanbag in their living room, cradling Jack in his lap with one hand. In the other, he held a syringe containing the nine-month-old's breakfast: 8 micrograms of sirolimus, a potent immune suppressant, prescribed to make sure his body didn't reject the gene therapy. Evan mimicked a cartoonish swigging sound and squirted it down Jack's throat.

Jack swallowed and pulled himself up by Evan's hand, a feat of infantine strength Naomi didn't achieve at that age. He never smiled before the treatment either. Now, he giggles.

"Is it the gene therapy? Is it just Jack being different? Is it the steroids?" Rebekah said. "But it's too many different things, in my opinion, to just be the steroids."

For the Lockards, parenthood was bliss. Then it was terror. Now it is something in between.

They still consider themselves lucky: Millions of children with ultra-rare genetic diseases—and their baby, almost alone, got treated. Rebekah is in a Facebook group for families with all forms of SPG. (There are over eighty.) But she's told almost no one about Jack's therapy. It would feel like a brag.

She also can't bear to look at the posts from parents of older children, struggling. They seem like omens of Naomi's possible future.

Since the therapy—a process that required flying to Dallas for two months on a moment's notice, leaving Naomi behind for most of it—both children have gained skills. Naomi lights up when she sees Evan. She started saying "Mama" and, finally, learned to crawl. They try to take joy in these moments. "She can now get in trouble," Rebekah said with a laugh.

Yet the Lockards know that, without treatment, she will decline. As she crawled up a ramp in the play area upstairs, you could see her ankles turn inward, an early sign of the spasticity that will eventually progress into paralysis. She was due to start three weeks of intensive occupational therapy. It cost $10,000 out of pocket.

Rebekah laughed off the expense, joking grimly they won't need to pay for college. They don't know what either child will need. Rebekah wants Naomi to just gain the cognition and motor skills required to communicate her needs and do small tasks, like cook with her mom. They know she will likely always be dependent. Jack is a black box.

"We'll see what the future holds," Rebekah said, tickling Jack's back as he played with a toy called the spinning rainbow. "You're a lucky boy, aren't you?"

Superhero or Don Quixote?

Naomi still hasn't been treated. Pirovolakis planned to start the clinical trial for her and other children with SPG50 in September with a $15 million grant from the California Institute for Regenerative Medicine, or CIRM.

Then CIRM paused all grant applications amid a leadership shuffle at the organization.

The Lockards and other families have been trying to raise money themselves, but the needs are huge: around $1.5 million. Treating Jack alone cost $450,000 in hospital and other costs. Waiting has taken its toll. Both parents are in therapy. Evan started medication. Rebekah is considering the same.

The plan is to now start a trial in March, but Pirovolakis acknowledges that's uncertain.

"It's a lot of worry," Rebekah said. "The treatment exists, it's ready to go, and every day that Naomi waits, she's losing neurons."

It's worn on Pirovolakis too, though he's remained upbeat. He'll cheerfully explain how close Elpida is to collapse. Despite his best efforts, investors didn't bite at his model. The biotech sector tanked. The diseases he targeted were too small to interest even philanthropic groups like Chan Zuckerberg and the Gates Foundation. GoFundMe campaigns bring in dregs. Only CIRM floated anywhere close to the money needed.

There are two ways to look at his story. There is the superhero narrative, this Tony Stark of biology, bending a broken medical system to bring cutting-edge technology to children who would otherwise be forgotten. Then there is the tragic story: Terry Pirovolakis as the proverbial child on the local news selling lemonade in a desperate but futile effort to raise money for his own cancer treatment, an indictment of everything wrong with modern health care, coated in the thin, feel-good veneer of a father saving his son.

"I think what he's doing is just extraordinary," said Frédéric Revah, CEO of the French gene therapy nonprofit Généthon. "But the whole field cannot depend on guys like Terry."

Pirovolakis shrugs when asked which lens he prefers. "I'd rather be an Amazon executive," he said. "But, you know, that's not how things work out."

Elpida is up to six programs now: two core ones and four unofficial ones that Pirovolakis won't commit to until he's secured more funding. He is still getting calls nonstop. More entrepreneurial parents like him are coming, he says. Once or twice a week, he advises a family on research agreements. He tries to keep Michael out of it. The boy, he figures, has been in the spotlight enough. But he's still a proud father. At a recent conference, he pulled out a phone and, beaming, showed people a video: It was six-year-old Michael getting on his bicycle and riding himself, undirected, to the beach.

TOM McALLISTER

When We Used to Glow

FROM *Hippocampus*

ALL SUMMER, WE'VE been talking about how there used to be more lightning bugs—some people call them fireflies, not the name I grew up with—and I'm worried that what we're really talking about is how we used to be younger, how we used to run around in the yard without worrying about the degenerative arthritis in our ankles, used to imprison those little miracles between clasped hands and stare in wonder as light pulsed in the gap between our fingers like E.T.'s heartbeat, used to free them sometimes but more often crush them in our palms because it was easy and because the other kids were doing it, used to smear the phosphorescent goo from their corpses across our cheekbones and our foreheads in consecration, used to run over to our parents all smiles and demand they witness us glowing like something out of Neverland, used to go back and capture more and more until at some point we were too tired to stand.

When we say there used to be more lightning bugs, I'm afraid we're really talking about how our parents used to be alive. I'm afraid we're saying there was a time when anything seemed possible, and now we can barely remember it. I'm afraid I'm on the verge of following all those Facebook pages that post a picture of chalk erasers with a caption like "If you know what these are you had a great life." I'm afraid if I let it go too far, I'll convince myself I did have a great life.

I remember summer nights looking out and seeing hundreds of them flashing in the trees. Sometimes one landed on my arm and I resisted the urge to swat it away, bathing in its light. Now, I can go a whole week without seeing even one. When I google "Are there fewer lightning bugs today," I get mixed results, but choose to relate the information that most hews to my beliefs. I tell my wife the internet confirms their population is in decline, but nobody is sure why. Pesticides maybe. Climate change maybe. Good enough guesses. My wife used to text her father early in the summer, whenever she saw the first lightning bug of the season. He would reply "Wonderful!" He's dead now, but she still feels that urge to text him. That's one of the ways you keep someone alive.

The seasons are not what they used to be either—summer is longer and hotter now, and we pinball between droughts and severe storms that sometimes last a week. New Jersey gets regular tornados between May and October. The rhythms of our lives are out of sync with the planet. In early August, we sat outside on a cool summer night (shorts, sweatshirt half-zipped) and talked about how every summer night used to be like this. It could be that this is just a new phase of life, where we yearn for a time that never really existed. But I think you can feel it too. It's not the same out there.

We don't use plastic wrap or Ziploc bags anymore. We buy reusable and recycled products whenever possible. We stopped serving guests on plastic; too many times we hauled out three full bags of trash with nothing but cups & plates & forks & knives that would outlive us all. Every couple months there's a new article about how recycling doesn't work, has never worked. We keep doing it, because it still seems preferable to the alternative. It is important to us to feel like we're being good people. I take a fish oil supplement every morning because my cholesterol is too high and getting higher. I want easy fixes. I want my choices to matter. Ideally, only the good ones.

A few years ago a friend from Portland traveled across the

country to visit, and we were sitting outside on our deck when some lightning bugs emerged in the night sky. My friend had never seen one in real life and, she later confessed, she'd been halfway convinced they were a mythical creature. Something that existed only to populate the background in cartoons; a figure you insert into a story to suggest the possibility of magic. When she saw her first lightning bug, she leaped out of her seat and ran down into the yard, running in circles and chasing them, calling out to us and laughing, exactly the way we used to do.

DAVID NAIMON

Eleven Stills

FROM *Prairie Schooner*

Results suggest that the share of the world's population travelling by air in 2018 was 11%, with at most 4% taking international flights.
—*Global Environmental Change*, Volume 65, November 2020

I

One of the oldest beings in Europe has stood in the same place for three thousand to five thousand years. Older than Ancient Greece. Legend has it that Pontius Pilate was born beneath its canopy. Wars, plagues, human and natural disasters alike, the Fortingall Yew stood stationary and still regardless. That is, until this very year, when mysteriously, this yew, a male tree, a producer of small cones, a pollinator, decided to be something else, someone else. No longer only a three-thousand-year-old he, he grew some red berries, and traveled to she.

2

To travel the world, to conquer lands in the name of God and country, to steal goods and make people property, one first needs to uproot trees. One 100-gun ship in Horatio Nelson's eighteenth-century British navy required four thousand oak trees, forty hectares of forest for a single vessel that sailed for merely twelve years. Nelson's own ship, the HMS *Victory*, required six thousand trees and the Royal Navy entire an estimated 1.2 million high-quality oaks. The Battle of San Juan, the Battle of Grand Turk, the Siege of Calvi, the Battle of Genoa, the Battle of Hyères Islands, the Battle of Cape St. Vincent, the Assault on Cádiz, the Battle of Santa Cruz de Tenerife, the Battle of the Nile, the Siege of Malta, the Battle of Copenhagen, the Raid on Boulogne—"I have ever been and shall die a firm friend to our present colonial system," declared Nelson, fittingly shot dead pacing the hardwood deck of the *Victory* in the Battle of Trafalgar.

3

The yews are not the only beings to reconsider. Orcas have been changing too. In 2020, White Gladis, the matriarch of an Iberian orca pod, exhibited behavior never before witnessed by humans. She circled small yachts and sailboats, rammed them, and ripped away their rudders. White Gladis was pregnant at the time, but shortly after giving birth, in contrast to all known orca behavior, she brought her new calf with her to continue her attacks. Orcas have only a small number of calves in a lifetime and are thus fiercely protective of them, often teaching them for years before allowing them to venture forth from their shelter. But White Gladis put the stopping of boats before questions of safety or family. And not long after, other orcas began to learn Gladis's techniques. This growing trend is still a mystery to scientists four years later—more than five hundred boat-orca incidents and counting, and now almost entirely carried out by juvenile orcas. Some boats, like the Swiss sailing yacht the *Champagne*, were sunk entire. The consensus advice for you, a fancy person on a fancy boat, if you are attacked by a killer whale, is to stop. Stop your boat. And leave the rudder loose. Don't move, don't go. Just stop.

4

The word "journey," at its root, has a constraint within it. One thousand years ago, the Old French word *journée* meant a day's length or a day's work or a day's travel. In Middle English, it meant the same thing, both a day and the distance traveled in a day. As recently as 250 years ago, at the dawn of the industrial revolution, it still held this sense. Travel was still constrained by the earth and its patterns. Even without movement, even standing still, the day itself has journeyed, and you have too, with it. Still.

5

One of the least-known no-fly zones in the world, where no plane or drone or other mechanized flight is allowed, is over Malabar Hill in Mumbai, India. Deep within this peacock-filled fifty-four-acre forest lies a Tower of Silence. And above this tower the skies are unobstructed by human activity. These circular raised structures have three concentric layers and are built in three ceremonial phases. The ground is dug, up to eleven feet deep, with prayers offered to the angel that guides souls on their journey beyond the body. The laying of the foundation begins with 301 nails driven into the earth. And one sturdy thread, which can travel the circumference of the future tower three times, is created by the intertwining of 101 fine threads, together threaded as one through the eyelets of the nails. This creates a spiritual web, network, bridge before the physical structure is made, one that will do the same between earth and sky, body and soul.

6

Each morning, commonly if mysteriously, vultures stand still in direct sunshine, their backs to the sun, their giant wings spread and held open in what is called a "horaltic pose." Some say the word "horaltic" derives from a mishearing of "heraldic"; others say it relates to "horal" or "by the hour"; and still others that it is a nod to the falcon-headed god of Egypt, Horus. Equally unknown is why vultures assume this stationary sunlit pose each day. There are many theories, from killing parasites in their wings to aiding with digestion. Some birds, like the hummingbird, can go into full torpor at night, dropping their temperature by as much as 50 degrees to conserve energy, making them seem unnaturally still, lifeless, and unresponsive. Vultures go into a mini-torpor, and some say the sunbathing pose is part of their revival each day, the return of life with the return of the day.

7

Towers of Silence, a Zoroastrian tradition many millennia old, as old as the Fortingall Yew, are sites of excarnation, or "sky burial." The corpses of men are placed in the outer circle, of women in the middle circle, and of children in the inner circle, whose axis is a central pit where bones are piled once the bodies have been stripped of their flesh. Vultures are the medium and the means for the soul to journey from its body, arriving to these raised towers at the high point of a hill, towers designed specifically to welcome them, to assist humans as they cross over. The sky above is full, not of humans in planes but of birds and souls.

8

The oldest form of human art, the cave art of ancient humans, the beginnings of figurative art, rarely depicted humans. And some of this art was not made by humans at all but by Neanderthals, a species, like so many others, we long presumed were incapable of complex thought or sophisticated belief systems. The oldest known cave art, forty-four thousand years old, contains therianthropes, creatures that are half-animal and half-human, therianthropes hunting warty pigs and dwarf buffalo. To scientists this suggests that the earliest human art, created by so-called primitive peoples, contained all the key elements of what we consider modern cognition—figuration and narration—the ability for abstract thought and imagination.

9

The science fiction and fantasy writer Ursula K. Le Guin suggests that until very recently, all great literature was imaginative and fantastical, whether *The Odyssey, Don Quixote, Beowulf, Hamlet,* or *The Metamorphosis.* That "realism in fiction is a recent literary invention, not much older than the steam engine and probably related to it." Le Guin also suggests that works of realism move toward anthropocentrism and imaginative literature away.

10

Not long ago, in our real world, there were forty million vultures in the skies of the Indian subcontinent. Today, with a 99 percent decline, the skies defy augury. The Temple of Silence is now not scattered with cleaned-over bones from bird-assisted soul travel but corpses in various states of putrefaction. Now far too few vultures exist to release human souls from their bodies. And the increasing smell in the downwind neighborhoods has divided the community on whether to abandon this many-millennia-old ritual and in what way.

11

The earth, the water, the air, the yew, the orca, the vulture, there is no way out but in, they say. You must root down and return to the limits of your body. The language that tethers humans to the earth, Proto-Indo-European, was spoken at the sprouting of the Fortingall Yew, during the construction of the first Tower of Silence. The root word shared by both the word "human" and the word "humus" (for earth, for soil) was spoken only in this now-extinct language. What would a journey to this word look like, sound like? Would we have to assume antlers or look through the eyes of a bird? How would or wouldn't we move? And where wouldn't we go?

EMILY RABOTEAU

Gutbucket

FROM *Orion*

I am a mother raising Black children in New York City, which is unceded Munsee Lenape territory. Often, I am afraid for my children's lives. Where my family lives, the storms are growing worse, and the water is rising, and these are not the only threats to our safety. I have come to the Arctic to ask you what changes you have witnessed, and to humbly ask, with your permission, for your wisdom about survival.

THIS WAS THE script I had rehearsed for my journey to a coastal Alaskan village in the late summer of 2022, when my boys were eleven and nine. It felt like something I would have to revise, potentially extractive and teetering on false equivalence, yet important to get right to justify the carbon footprint of traveling such lengths and leaving my children behind. The last place I'd flown this far from home to study survival was Palestine.

I was joining my colleague Dr. Maria Tzortziou, a distinguished professor of environmental sciences who studies the effects of climate change upon vulnerable ecosystems and communities, including the basin where the Yukon River joins the Bering Sea. This is one of the fastest warming parts of the planet, or as Maria put it, "a ground zero." Ours was an interdisciplinary effort.

While she prepared to conduct measurements with her NASA-funded research team that would capture changes in the coastline from as far as space, we'd work together gathering testimony from elders in the Native community whose tribal council office would serve as our base, having collaborated with Maria's team for several years. To truly understand the impacts of environmental change in the coastal Arctic, Maria explained, scientific questions need to derive from the people who live there. For instance: Why are the salmon dying?

To say I felt like I was traveling to the edge of Earth only indicates my bias. Technically, I wasn't even leaving the United States. To the Yup'ik people who've lived through subsistence hunting and gathering in the Yukon-Kuskokwim Delta for millennia, that region is the center. It took two days and five flights to get there, each plane smaller and later than the one preceding it:

New York to Chicago.

Chicago to Anchorage.

Anchorage to Bethel.

Bethel to Emmonak.

Emmonak to Alakanuk.

On the second of these flights, the old man seated directly behind me suffered from severe memory loss. Every few minutes, for six straight hours, he asked his companion, whom I took to be his long-suffering wife, where they were going. "Anchorage, Rafael," she repeated. "How many times do I have to tell you?"

Three minutes later, he'd ask her again: "Where are we going?"

"I already told you," she'd moan. "You're not connecting the dots."

Adding to the sense of disorientation (mine) were the pink drifts of fog lit up by the low-hanging sun, not yet set though it was late at night, because we were so far north.

On the fourth of these flights, a tiny biplane carrying more weight in mail than human cargo, I found myself buckled next to a woman in her fifties crying like an animal in a trap. She had no companion to ground her. We were about to take off. Her sobs racked her body and filled the aircraft. I knew better than to ask if she was okay, because she so clearly wasn't. Her pain was elemental, naked, pressed up against me, unconscionable to ignore. Not since my friend Drema, keening over the loss of her dad to Covid-19 on her front stoop in Harlem, had I heard such a sound. The difference then was that we weren't allowed to touch each other to offer comfort. I could freely offer my seatmate comfort through touch, but I hesitated, thrown off by jet lag, the weight of the moment, the weight of the era. What day was it? Where was I? What to do? What to say? Who was this woman's tribe? Why was she mourning? Would she welcome a stranger's intervention or prefer to be left alone? Then I remembered a piece of advice I'd been given by my mother. If there's ever a question of whether to offer kindness to someone in pain, the answer is yes. Do it.

I laid my hand on her heaving shoulder and pressed a tissue into her clenched fist. The engine kicked up. The propellers whirred. The woman, whose name was Edna, turned her wet, pained face to mine and choked, "I just heard my mother died. I'm going to Emmonak to see her. But I'm too late. Why do I have this feeling that I'm all alone? Would it be all right if you hold me?"

I knew exactly how to hold her because I was a mother. Because I had been held. I stroked the side of her face, her arms, as if she were my baby. *Sssshhhh*, I said. The earth below was not solid, was all pocked with puddles and ponds. The delta. For the rest of this flight, I now understood, my role was to orient myself to her suffering. We needed a ritual to deliver us. I unwound the bright pink kaffiyeh from my neck and put it around Edna's shoulders. I'd gotten it in Palestine. It wasn't a quilt, but it would have to do. "I want you to have this because your heart is in a dark place," I told her.

"This cloth is to remind you that there is still color in the world, and when you can see it again, you will give it away to someone else."

"*Quyana*," Edna said, using one corner of the scarf to wipe her eyes. Somehow, I knew the word meant "thanks." "I really needed the support." Presently, her breath calmed down. We sat together in our travail until landing. Instead of saying goodbye, she asked, "It's never going to be the same again, is it?"

"No," I said. I told Edna I was sorry. I told her I thought I knew how she felt. "I lost my father a year ago. You feel like an orphan."

The singing style of the slaves, which was influenced by their African heritage, was characterized by a strong emphasis on call and response, polyrhythms, syncopation, ornamentation, slides from one note to another, and repetition. Other stylistic features included body movement, hand-clapping, foot-tapping, and heterophony. This African style of song performance could not be reduced to musical notations, which explains why printed versions do not capture the peculiar flavor of the slave songs . . .

—Albert Raboteau, *Slave Religion*

Let me tell you what, in his lifetime, my father loved. He was always on the lookout for thin places, where the distance between heaven and earth collapses so we're able to glimpse the divine. My father collected icons, attempts to render the human face of God. He loved the spiritual power of the Divine Liturgy. He loved the teachings of Trappist monk, mystic, theologian, and social activist Thomas Merton, who said: "Life is this simple: we are living in a world that is absolutely transparent and the divine is shining through it all the time." He loved his children. My brothers. Me. His grandchildren. My boys.

One area of his research as a renowned historian of African American religion was the retentions, or "Africanisms," that survived the lacuna of the Middle Passage as expressions of faith among the enslaved even as they took on, or were made to take on, the master's religion. The blue note in spirituals. The dance in the ring shout. The syncretism of a Yoruba deity with a Catholic saint. The rituals that protect a soul's core from getting snuffed out by evil forces putting profit above life. This was the stuff that interested my father.

At his seventy-seventh birthday party, he took down the framed photos from what he called his "ancestor wall." He was well into his diagnosis by then and had started hospice care. His hands were trembling. My stepmother had baked a German chocolate cake. The pecans in the icing reminded him of the Mississippi Delta, his home place, in the eastern floodplain where the Yazoo and Mississippi Rivers join the Gulf of Mexico, increasingly a dead zone where fish are dying, though it's just as important for you to know that this mucky region was also the birthplace of the blues. He showed us the pictures, my brothers and me and our many children who had gathered to celebrate his life, which was ending. As he named our ancestors for us and reminded us of their stories, I thought—he knows he is going to forget them. I understood to pay keen attention, lest the names be lost for good.

> *Agnes Damian, Arthur Chikigak, Donald Augustine, Frank Alstrom Jr., Nellie Edmund, Ragnar Alstrom, Raymond Oney, Regis Augline, Richard Agayar, Rose Isidore, Ruth Oney, Sally Leopold, Mary Ayunerak, Denis and Winifred Sheldon . . .*
> —List of elders, tribal council office, Alakanuk,
> August 2022

The fifth and final flight did not include our luggage and so Maria and I arrived in Alakanuk with nothing but our questions. Maria was understandably miffed about the delay of her field equipment, but glad to be back in what she described as a "special place." I was in a state of awe, blown away by the beauty of the wetlands—not water, not earth, but both.

A man called Theodore Hamilton was there at the runway to pick us up. "Welcome to the Yukon," he greeted me. He gave Maria a big bear hug, pleased to see her again. She was last in Alakanuk in spring, when she started training local high school students to collect environmental samples. We climbed onto the back of his red four-wheeler. Theodore could talk, smoke, and hunt. He was a self-described "half-breed," and by the time he drove us through low willow thickets along the muddy road to the elevated one-story plywood-sided cabin of his boss, Augusta Edmond, he'd already explained in as elegant a disquisition on ambiguous loss and traditional ecological knowledge as I'm sure I'll ever hear: that he sees through Yup'ik eyes, that the land is his church with no walls, that he's an outlaw in the eyes of state bureaucrats who try to impose regulations through white man's laws thinking they know best when it's local communities who know how to take care of the land sustainably, that his people used to hunt in quadrants in rotation across the domain to keep regeneration in balance following the seasons and the animals, that in camp, on the tundra, you have to wash town off you and get quiet, that a September moose tastes better than an August moose, that there are more moose now, that the willow trees they eat are getting taller now that there's not enough snow, that the snow geese and cranes are the smartest birds, that there are fewer ducks now in the sloughs than before, and fewer porcupines too, that you never used to see a beaver when he was a kid but now that the terrain is different they're all over the place, that the fat of the black bear turns blue after the thaw when they grow drunk from eating so many blackberries, that the speck in the sky over there is a loon, that down in the lower forty-eight where they paved over the wetlands they think they know better, that the white man's food makes you sick and the white man's ways have poisoned the salmon, that from the headwaters to the mouth, nobody is fishing for salmon because no salmon are passing through, that things are gonna get harder, that the spring freshet used to feel like an earthquake in his heart when the ice came downriver, but not no more.

He parked in a mud puddle. We were at Augusta's place. Augusta had been managing the tribes' environmental protection programs for many years, with Theodore as her assistant. Like a lot of Bureau of Indian Affairs houses in that village of roughly six hundred, on account of permafrost melt, Augusta's place was shifting. (Some homes had to be moved away from the river. Because of rapid erosion they were at risk of falling in.) And like a lot of working mothers, including me, her house was cluttered. Yet she opened her home to us as if we were family. She pulled two pairs of sweatpants out of an enormous pile of clean laundry yet to be folded for Maria and me to use as pajamas until our bags turned up. Maria asked after her kids. The sweatpants belonged to Augusta's daughters. She had five daughters, she told us later when I asked what she feared and hoped for her children, and though it worried her to imagine—along with loss of subsistence hunting ways, the loss of the sea ice, the loss of the permafrost, the loss of the snow, and the loss of the salmon—she understood why young people might not want to stay in the village, because the village would inevitably have to move to firmer ground. But Augusta also said she was busy, she had things to do; there were two hundred boxes of meat from the Food Bank of America on the way, they would have to be distributed, so if I really wanted to know about climate change, I should talk to the elders, who remembered how it used to be before the warming began.

> *Death is someone you see very clearly with eyes in the center of your heart: eyes that see not by reacting to light, but by reacting to a kind of a chill from within the marrow of your own life.*
> —Thomas Merton, *The Seven Storey Mountain*

After my father died, a friend introduced me to the term "ambiguous loss." It seems important to mention that this friend was preoccupied by the moral question posed by the climate emergency of bringing children into this world. Ambiguous loss, my friend explained, was a theory introduced in the 1970s by social scientist Pauline Boss while researching the experience of unresolved grief among families of soldiers gone missing in the Vietnam War. This kind of loss doesn't come with closure or clear understanding. When a loved one goes physically or mentally missing, it leaves the aggrieved unsettled, unresolved, searching for answers. It can occur, as in families scarred by the Holocaust or slavery, across generations. The therapeutic guidance for such trauma is toward building resilience. Other forms of ambiguous loss might result from incarceration, migration, divorce, miscarriage, terrorism, addiction, pandemic, climate chaos, or dementia, like my dad's.

Indeed, by his seventy-eighth birthday party, my father had forgotten the names of our forebears. And not only their names: He had also forgotten his own name. The names of his children and grandchildren. How to feed himself. How to dress. How to write. How to read. One of the last things he remembered was the first eighteen lines of *The Canterbury Tales*, which he could mouth along with my recitation, according to something like the muscle memory of that aged ballerina whose Alzheimer's didn't rob her of the choreography when *Swan Lake* was played. *Whan that Aprill with his shoures soote / The droghte of March hath perced to the roote, / And bathed every veyne in swich licour.* . . . And then, he forgot that too.

He was not yet dead but the lambent mind I had known was gone. Did he know where he was going? "It's okay, Dad," I told him, reintroducing him to old photos one by one. "This is your mother, Mabel, who raised you to overcome. She brought you north to keep you safe. She went gray young and didn't suffer fools. She took you to church every Sunday. That is you, when you were her little boy, sitting on her lap. This is her husband, your father, Albert, standing between your big sisters, Marlene and Alice. She named you after him. This is her favorite sister, your aunt Emily. You named me after her. This is her grandfather, Edward, whose mother, Mary Lloyd, was enslaved . . ." These are our people, I meant to say, from whom we come and to whom we belong.

> *The flexible, improvisational structure of the spirituals gave them the capacity to fit an individual slave's specific experience into the consciousness of the group. One person's sorrow or joy became everyone's, through song. Singing the spirituals was therefore both an intensely personal and vividly communal experience in which an individual received consolation for sorrow and gained a heightening of joy because his experience was shared.*
>
> —Albert Raboteau, *Slave Religion*

On Sunday I walked from the tribal council to Alakanuk's Catholic church. The mud sucked at my hiking boots. It was the twenty-second Sunday in Ordinary Time. Ordinary time? In two days, I'd overhear Maria reading aloud to her team an article from the Associated Press stating that "dead ice" from the rapidly melting Greenland ice sheet would eventually raise global sea levels by ten inches. We were eating a breakfast of instant oatmeal at the bustling tribal council office, having rolled up our sleeping bags so the researchers could share this vibrant space with the people employed to serve the community. "It's just going to melt and disappear from the ice sheet," she read. "This ice has been consigned to the ocean, regardless of what climate (emissions) scenario we take now." Later in the article, a glaciologist at the Greenland survey described the situation as "one foot in the grave."

"Does that mean I can stop recycling?" cracked one of the scientists, deploying gallows humor before gamely going out into the bad weather, despite the grim prognosis, to collect samples in the boat of a local called John Strongheart. Meanwhile, Augusta and Theodore were figuring out how best to get the government meat to the folks in town who could no longer live off hunting and gathering alone. The ancestors on the wall vigilantly watched our true acts of devotion.

Nearby the church stood a tide staff for measuring flood events. The high-water mark was eleven and a half feet. The recommended building elevation was a foot higher than that. The church, like the tribal council, was elevated on stilts. Jesus hung on the cross at the altar. Before the cross, wearing green robes and rain boots, stood the deacon, Denis Sheldon. He had a stooped back and a kind face. The acolyte passing out missalette hymnals was his wife, Winifred. One of the hymnals was in Yup'ik. The other hymnal was in English. In a front pew knelt Mary Ayunerak, wearing a flower-print headscarf knotted below her chin. Like Denis and Winifred Sheldon, she was an elder. I knelt beside her. Maybe because it was the start of moose hunting season, the church was otherwise empty.

The reading was from the Book of Sirach, on humility. How to live within the covenant, faithful to God in the small things. "We know this lesson is true," Denis Sheldon confirmed in his sermon, "because we were already told by our elders to be humble. I'm glad I was born when I was, because in my childhood I was taught by elders with knowledge before the missionaries came." On the wall behind the cross hung a fishnet with a harpoon, two eagle feather dance fans, animal skins, and a drum. Clearly, the cross had been absorbed into a larger cosmology. Jesus was only a small part of the story. Denis, Winifred, and Mary sang a song in Yup'ik. Toward the end of the Mass, I joined my voice with theirs to sing "Go, Be Justice." I remembered the hymn from my Catholic upbringing for this rousing lyric: "Catch the tyrants in their lies." I liked singing that song with this congregation. When the service was over, I told them why I had come. They agreed to speak with me and Maria, back at the tribal council, the following day.

Mary Ayunerak, who has sixteen grandchildren and twelve great-grandchildren, and whose phone number was listed on a suicide prevention poster as someone in the community available to call on for encouragement and support in times of trial, said that the Yukon didn't used to be that wide. Her original birthplace, close by, is now underwater. "Right now, our water is dirty," she said, "it's so messed up, it never used to be like that. It wasn't that warm. Used to be kind of cool. Really nice to drink out of too, from the river. We didn't used to have sandbars down here. A lot of homes that used to be down here are gone due to the eroding. We used to get a lot of fish, wherever we set net, but now, no. We used to work together as a community to get food for the winter supply, and whatever we need, but now . . . people hardly ever go fish camping because no fish to cut. We don't go anywhere anymore like we used to a long time ago. Fish are starting to get pus-y and they have that ugly smell, and so maybe that's why people don't ever hardly go out as a family anymore. We used to be able to go out and get whatever we need and eat together, connect things together. Our homes are sinking. The ground is getting soft. Our permafrost is not like it used to be long ago. We used to travel a lot and put our food that we catch under the ground in a little house-like thing where we kept our food nice and cool, but the food started to smell."

Interested in learning how the younger generation was dealing with these changes, Maria asked if her grandkids worry. Mary Ayunerak said, "Yeah, my grandkids do." One of them, she couldn't remember who, asked her, "Grandma, where do you think we'll move, like if we have to move?" And she told them, "I don't know, I might not be here, and if I happen to be here, I'll be too senile to know where we are. It will be up to you guys to decide."

Winifred Sheldon picked up Mary Ayunerak's thread on biodiversity loss: "We had some dead fish floating on the Yukon. We're having a hard time getting the fish, and the fish is very important to us. We eat that all winter right now. And right now, it's getting harder. It really is changing a whole lot compared to when we were small. My mom used to sew lots of boots when we were young with beaver skins for the bottom, and spotted seals for the top. I almost learned how to make the boots, but I went to high school almost five hundred miles from here and then I forgot how to do it. I had to stay with white people, and I felt uncomfortable at times because I was the only Yup'ik person there, and sometimes I would feel like, 'Why was I born Yup'ik?' when I was staying with them. After two years, I came home, and I never went back. I wish

I had just stayed home and learned how to do things like my mom did, like sewing. I would have learned to make those boots for my grandkids or my kids when they were small."

Denis Sheldon shared his real name before continuing his wife's story about what had been lost. Kituuralria. He told me it means "the one who's passing by."

"It was a time when the whole country here was tundra," he said, "and caribou used to come. During the fall time, they would herd the caribou toward the river, that's when they killed what they wanted. Most of the time it was young calves because calves made the best clothing.

"There's been a lot of changes in the sloughs and channels. Beluga whales are not as many. Some seals are not as many as they used to be. Like the spotted seals. Right now, the freeze-ups are getting later and later, and breakups are getting to be earlier and earlier. When we were young, our men did a lot of trapping for mink, otters, muskrats, foxes. In those days, when it was colder, the men would be able to cross the Yukon with their dog teams. They could see the frozen breath of the dogs in the air like smoke.

"When I see pictures like that up there," he said, looking up at the ancestor wall of photos of the dead, some of their names passed down and echoed in the list of elders hanging by the phone, such that the tribal council office felt like a genealogy, like reading Leviticus, "they're all people who took care of the land, the water. Like that man to the right toward the window with red suspenders—Joe Phillip."

"That's my father," explained Winifred Sheldon.

"I learned a lot from him," Denis Sheldon continued. "He would say, 'Show respect to the animals.' They would have fish in ceremonies and rites they did because they believed that everything has a spirit. Yeah, so our people were very spiritual. The one that they respected most is the Great Spirit of the Universe. They always told us, wherever you go, even if you're all alone, you're not alone, because the one who is not seen is there with you. They had many names for Him, the Great Spirit. The Almighty. There were so many names they gave him. So many names. The greatest command was to get to know what He wants us to be."

I told Denis Sheldon that I liked the sermon about humility he gave on Sunday and asked if he'd been sent to a Catholic residential school as a child, like his wife.

"Yeah," he said, "close to one hundred miles from home." He was eleven, the same age as my eldest, when he went away to that school. He was nineteen when he came back. He lamented, "I really missed my family when I went. Those who took care of us, some of them were not good. They mistreated us. Some of those effects . . . I can see how they affected me. I tried not to show it."

I asked if he felt the pope's recent apology for the church's role in the cultural genocide that went down in the state-sponsored residential schools with mottos like, "Kill the Indian in the child"—where children like Denis and his wife, Winifred, were torn from their families, had their hair shorn, were punished for speaking their Native languages, and in many cases were psychologically, spiritually, physically, sexually abused, and even murdered, all in the name of being "civilized"—was meaningful. I didn't ask it like that, though. I simply asked if he accepted the pope's apology. He said yes, but also that the apology wasn't enough. In the deacon's view, the expression of sorrow needed to be spread to those who didn't hear it. "Our children and grandchildren," he said, "I think they need to know that some things happened that were not good. That affected their lives somehow." That was how Denis Sheldon spoke about generational trauma.

"How do you reconcile the need for justice with the need for forgiveness?" I asked him. "How does your body bear it?" This was what I had come all this way to learn—*What do we do with our anger? What can we do when we can't just stop living?*

"That's easy." The deacon smiled. His eyes had been smiling all along. Coming from someone else's mouth, his answer may have sounded trite. But not when Denis Sheldon said it. I felt completely disarmed by Kituuralria's *qanruyutet*, knowing in my bones these words of wisdom to be true: "We take care of each other."

How do you transform and transfigure sorrow into joy? That's the theme of the greatest Christian poet in the English language of the 19th century, Elizabeth Barrett Browning. There's no accident that Du Bois chooses her twice to invoke the epigraphs of The Souls of Black Folk. "What is it, Elizabeth?" "I'm trying to saturate Christ's blood in my soul so that I can respond to the ceaseless wailing of suffering humanity and transform sorrow into joy." It's no accident that he would be the one to get inside the humanity of Black folk in white supremacist USA. From Bay St. Louis, Mississippi—Gutbucket, Jim Crow, Mississippi. Who would think that after the murder of his father he decided not to hate, not to terrorize, not to traumatize. He's gonna be a love warrior, freedom fighter, yes. That's what he was, but with a deep sense of humility. Deep sense of self-criticism. Oftentimes I would say, "Brother, you're too hard on yourself." I'm Holy Ghost Baptist so I don't have to worry about that. We Baptists don't have to worry about being too hard on ourselves. We need to be more humble. Brother Raboteau going through the rich Catholic tradition and went on his way to Russian Orthodoxy—condemnation of no one, absolute forgiveness of everyone, embracing humanity of everyone but always understanding that you look at the world through the lens of the cross—that's what made my brother a saint in my eyes, 'cause for me a saint ain't nothing but a sinner who looks at the world through the lens of the heart. And for Christians who look at the world through the lens of the cross.

—Cornel West at Al Raboteau's funeral,
September 24, 2021

"And this is your great-great-grandmother Philomena Laneaux," I told my children seated in the pews at my dad's homegoing, holding up the last of the many pictures I'd brought to the altar. My inheritance. (They will go on the walls of my office now.) I didn't want them to be afraid of the corpse, but rather to understand that he was now an ancestor too. The choir waited up in the loft of Mother of God, Joy of All Who Sorrow Church, ready to start the next hymn in a minor key. The icons surrounded us, their faces aglow. I felt a sense of communion, an old pattern of meaning I rarely had the distance to see; a web of relations spreading behind and before me. Soon we would gather round the casket, and then the grave. I finished my eulogy with a quote from Birago Diop's "Sighs," translated from the French, epigraph to *Slave Religion*.

Those who are dead are never gone:
they are there in the thickening shadow . . .

SCOTT SAYARE

The Smell Test

FROM *The New York Times Magazine*

AS A BOY, Les Milne carried an air of triumph about him, and an air of sorrow. Les was a particularly promising and energetic young man, an all-Scottish swim champion, head boy at his academy in Dundee, a top student bound for medical school. But when he was young, his father died; his mother was institutionalized with a diagnosis of manic depression, and he and his younger brother were effectively left to fend for themselves. His high school girlfriend, Joy, was drawn to him as much by his sadness as his talents, by his yearning for her care. "We were very, very much in love," Joy, now a flaxen-haired seventy-two-year-old grandmother, told me recently. In a somewhat less conventional way, she also adored the way Les smelled, and this aroma of salt and musk, accented with a suggestion of leather from the carbolic soap he used at the pool, formed for her a lasting sense of who he was. "It was just him," Joy said, a steadfast marker of his identity, no less distinctive than his face, his voice, his particular quality of mind.

Joy's had always been an unusually sensitive nose, the inheritance, she believes, of her maternal line. Her grandmother was a "hyperosmic," and she encouraged Joy, as a child, to make the most of her abilities, quizzing her on different varieties of rose, teaching her to distinguish the scent of the petals from the scent of the leaves from the scent of the pistils and stamens. Still, her grandmother did not think odor of any kind to be a polite topic of conversation, and however rich and enjoyable and dense with information the olfactory world might be, she urged her granddaughter to keep her experience of it to herself. Les only learned of Joy's peculiar nose well after their relationship began, on a trip to the Scandinavian far north. Joy would not stop going on about the creamy odor of the tundra, or what she insisted was the aroma of the cold itself.

Joy planned to go off to university in Paris or Rome. Faced with the prospect of tending to his mother alone, however, Les begged her to stay in Scotland. He trained as a doctor, she as a nurse; they married during his residency. He was soon the sort of capable young physician one might hope to meet, a practitioner of uncommon enthusiasm, and shortly after his thirtieth birthday, he was appointed consultant anesthesiologist at Macclesfield District General Hospital, outside Manchester, in England, the first in his graduating class to make consultant.

The Milnes installed themselves in an ancient stone farmhouse high on a country hill in Cheshire. By then they had three young sons, and the edifice, which was old enough to be listed in the Domesday Book of 1086, was a happy, never-ending project. They threw elaborate, boozy dinner parties; they kept geese and hens and took in stray cats, dogs, a duck. "We just seemed to get on and do things," Joy told me. Friends still liken her to Mary Poppins, part twinkly magic, part no-frills practicality. She considers herself to be a "never-stop person," she said. Her husband was the same.

Les spent long hours in the surgical theater, which in Macclesfield had little in the way of ventilation, and Joy typically found that he came home smelling of anesthetics, antiseptics, and blood. But he returned one August evening in 1982, shortly after his thirty-second birthday, smelling of something new and distinctly unsavory, of some thick must. From then on, the odor never ceased, though neither Les nor almost anyone but his wife could detect it. For Joy, even a small shift in her husband's aroma might have been cause for distress, but his scent now seemed to have changed fundamentally, as if replaced by that of someone else. She thought he smelled vaguely of his mother.

Les had lately begun to change in other ways, however, and soon the smell came to seem almost trivial. It was as if his personality had shifted. Les had rather suddenly become detached, ill-tempered, apathetic. He ceased helping out with many household chores; he snapped at his boys. At parties out, he took to behaving in an excessive, theatrically buoyant way, before returning home to collapse in exhaustion. Joy felt these performances were meant in part to spite her, to prove that her complaints about his new demeanor were unfounded. "There was no empathy," she said.

An odd physical deterioration also seemed to have begun. At the house, he was awkward with the mower. On a vacation to the Scottish coast, he and his sons took sailing lessons; the boys took to it quickly, but Les could not work out the timing. A golf club slipped from his hands midswing. Playing darts at a friend's house, he somehow managed to throw one into his foot.

Les's sleep, too, was increasingly disturbed, and bizarrely, he began acting out his dreams. He once dreamed of catching a burglar; Joy lied to friends and colleagues about the source of the bruises she was left with, thinking no one would believe that her husband had attacked her as he slept. Nor, at Les's unequivocal demand, did she speak a word of his newfound impotence. In private, he laid the blame on her. She had grown too fat, he said, to be of any possible interest.

The man Joy married had no such cruelty in him. She considered leaving. After more than a decade of alienation and resentment, however, it began to occur to her that the changes in her husband might have some organic cause, that they might be symptoms of some disease. When he began referring to "the other person," a shadow off to his side, she suspected a brain tumor. Eventually she prevailed upon him to see his doctor, who referred him to a neurologist in Manchester.

Parkinson's disease is typically classed as a movement disorder, and its most familiar symptoms—tremor, rigidity, a slowing known as bradykinesia—are indeed motoric. But the disease's autonomic, psychological, and cognitive symptoms are no less terrible and commonly begin during the so-called prodrome, years before any changes in movement. And yet they do not suggest a diagnosis. Irritability, fatigue, troubled sleep—these are extremely common among the well, and among the infirm they can be associated with any number of conditions; in Parkinson's, it is only in retrospect that they are revealed to have been the first trace of the illness. For Les, Joy realized, the symptoms had begun nearly a decade and a half before he saw a doctor. Had it only been possible then to see them for what they were, she thought, to reach a diagnosis when they first emerged, so much confusion, so much resentment, and so much pain might have been averted.

Of the truly devastating diseases of neurodegeneration, Parkinson's is, after Alzheimer's, the most prevalent. It is also said to be among the most treatable. The existing therapies address its symptoms alone, however. There is no cure, no intervention to halt the progression of the disease or reverse its damage to the brain. The great difficulty of early diagnosis is perhaps the principal hurdle. Parkinson's disease is still diagnosed essentially as it was two hundred years ago, on the basis of its characteristic motor symptoms. Yet by the time these emerge, most of the neurons it will kill have already died. "Trying to rescue the brain from that level of damage is very difficult," says David T. Dexter, a neuropharmacologist who oversees research at the medical foundation Parkinson's UK. "You're trying to close the stable door after the horse has bolted, really." For a disease-modifying therapy to have a meaningful shot at efficacy, it would most likely need to be applied far earlier, years or even decades before most patients are currently diagnosed. Tilo Kunath, a professor at the University of Edinburgh, told me that an early diagnostic was "a sort of holy grail" for the field, the prerequisite for almost everything else. The Michael J. Fox Foundation for Parkinson's Research, the world's largest nonprofit funder of Parkinson's science, has spent more than $900 million looking for such a biomarker. With the exception of one that involves a spinal tap, which considerably limits its utility, almost nothing has shown promise.

Les was forty-four. He began treatment with L-DOPA, the amino acid that has now been the standard of care in Parkinson's disease for sixty years. L-DOPA is extremely well tolerated, but over time its efficacy fluctuates and wanes. At a certain point the dopamine agonist ropinirole was added to Les's regimen. It was reasonably effective, but also, as Joy put it to me, "vile." Later she came to believe it was the cause of much of the inappropriate behavior her husband soon began to display; at the time, she had no explanation.

Dopamine agonists are now recognized to cause severe problems of impulse control, and Parkinson's patients often develop addictions to shopping or gambling. In Les, the problems were in the realm of sexuality. He stockpiled pornography and boasted of his sexual prowess to appalled younger women. At a party at the house in Cheshire shortly after his retirement, at fifty—his fatigue, in particular, had become unmanageable—Les came down the stairs in tartan-patterned underwear, sat in the lap of a female guest, and turned to the onlookers in glee.

The symptoms of the disease itself worsened too. Les's voice quieted, and he developed a slow, shuffling gait, often freezing midstride. Soon he could no longer dress himself; he began wearing diapers; his sleep was ever more irregular. The effects of his medication were also less and less predictable, and he seemed to flip randomly from relatively functional "on" periods to "off" periods of uncontrollable dyskinesia and despair.

Alzheimer's is often called the "long goodbye," but the term is a no-less-accurate description of Parkinson's, in which patients bear the added pain of self-awareness, of recognizing that a gulf has opened between themselves and the kingdom of the well, and of watching it grow wider each day. A friend of Joy's, Catherine Havranek, had a husband, Ivan, who likewise had Parkinson's. Ivan was a devoted amateur cellist. He chose to turn in his cherished instrument while he was still capable of carrying it up the stairs to the dealer on his own. "He said that it was too good a cello to waste on him," Catherine told me.

Joy's life was given over almost entirely to Les's care, as well as to the care of his chaotic mother, Helen. "When Les was diagnosed, she really went off the deep end," Joy told me. Helen was once again placed in a psychiatric ward, and she then moved in with Joy and Les. Soon thereafter, Helen herself was diagnosed with Parkinson's. In retrospect, it seemed almost certain that her longstanding psychiatric problems were in fact manifestations of the disease, misread for decades by doctors who conceived of Parkinson's as a movement disorder alone, and one erroneously thought to be quite rare in women.

After Helen's death, Joy and Les returned to Scotland, settling in Perth, a town in the Midlands. They were ever more estranged from each other, locked in their tense routine. They knew no one. Feeling desperate, Joy eventually persuaded Les to go with her to a meeting of local Parkinson's patients and their caregivers.

The room was half-full by the time they arrived. Near the coat stand, Joy squeezed behind a man just as he was taking off his jacket and suddenly felt a twitch in her neck, as if some fight-or-flight instinct had been activated, and she raised her nostrils instinctively to the air. She often had this reaction to strong, un-expected scents. In this case, bizarrely, it was the disagreeable odor that had hung about her husband for the past twenty-five years. The man smelled just like him, Joy realized. So too did all the other patients. The implications struck her immediately.

For nearly all the recorded history of medicine and until only quite recently, smell was a central preoccupation. The "miasma" theory of disease, predominant until the end of the nineteenth century, held that illnesses of all kinds were spread by noxious odors. By a similar token, particular scents were understood to be curative or prophylactic. More than anything, however, odor was a tool of diagnosis.

The ancients of Greece and China confirmed tuberculosis by tossing a patient's sputum onto hot coals and smelling the fumes. Typhoid fever has long been known to smell of baking bread; yellow fever smells of raw meat. The metabolic disorder phenylketonuria was discovered by way of the musty smell it leaves in urine, while fish-odor syndrome, or trimethylaminuria, is named for its scent. Dogs are able to smell an extraordinary range of illnesses, including Covid-19, though they have been tested most extensively for cancers. In one famous case from 1989, reported in *The Lancet*, a woman in England noticed that her Doberman–border collie cross had begun sniffing intently at a small mole on her leg. After several months, the dog tried to bite it off. This caused the woman to see a doctor. It was a malignant melanoma.

Most diseases can be identified by methods more precise and ostensibly scientific than aroma, however, and we tend to treat odor in general as a sort of taboo. "A venerable intellectual tradition has associated olfaction with the primitive and the childish," writes Mark Jenner, a professor of history at the University of York. Modern doctors are trained to diagnose by inspection, palpation, percussion, and auscultation; "inhalation" is not on the list, and social norms would discourage it if it were.

During her time as a nurse, Joy had done it anyway, reflexively, and learned to detect the acetone breath that signaled an impending diabetic episode, the wet brown cardboard aroma of tuberculosis—"not wet white cardboard, because wet white cardboard smells completely different," she explained—or the rancidness of leukemia. The notion that Parkinson's might have a distinctive scent of its own had not occurred to her then, but when it did occur to her years later, it was hardly exotic.

She and Les worried that the normosmics of the world, unfamiliar with medical smells and disinclined to talk about odor in general, might not take her discovery very seriously. They searched for an open-minded scientist and after several weeks settled on Kunath, the Parkinson's researcher at the University of Edinburgh. In 2012, Joy attended a public talk he gave. During the question-and-answer session, she stood to ask, "Do people with Parkinson's smell different?" Kunath recalls. "I said, 'Do you mean, Do people with Parkinson's lose their sense of smell?'" (Smell loss is in fact a common early symptom of the disease.) "And she said: 'No, no, no. I mean, Do they smell different?' And I was just like, 'Uh, no.'" Joy went home. Kunath returned to his usual work.

Six months later, however, at the urging of a colleague who had once been impressed by cancer-sniffing dogs, Kunath found Joy's name and called her. She told him the story of Les's new smell. "I think if she'd told me that, as he got Parkinson's, he had a change in smell, or if it came afterwards, I probably wouldn't have followed up any more," Kunath told me. "But it's this idea that it was years *before*."

He called Perdita Barran, an analytical chemist, to ask what she made of Joy's claims. Barran suspected Joy was simply smelling the usual odor of the elderly and infirm and misattributing it to Parkinson's. "I knew, because we all know, that old people are more smelly than young people," says Barran, who is now a professor of mass spectrometry at the University of Manchester. Still, Barran was personally acquainted with the oddities of olfaction. Following a bike accident, she had for several years experienced various bizarre distortions to her own sense of smell. The idea that Joy might be capable of experiencing odors that no one else could did not strike her as entirely outlandish.

She and Kunath ran a small pilot study in Edinburgh. Through Parkinson's UK, they recruited twelve participants: six local Parkinson's patients and six healthy controls. Each participant was asked to wear a freshly laundered T-shirt for twenty-four hours. The worn shirts were then cut in half down the center, and each half was placed in its own sealed plastic bag. Kunath oversaw the testing. Joy smelled the T-shirt halves at random and rated the intensity of their Parkinsonian odor. "She would find a positive one, and would say, 'There—it's right there. Can you not smell it?'" Kunath recalled. Neither he nor the graduate student assisting him could smell a thing.

Kunath unblinded the results at the end of the day. "We were on a little bit of a high," he recalled. Not only had Joy correctly identified each sample belonging to a Parkinson's patient, but she was also able, by smell, to match each sample half to its partner. Barran's skepticism evaporated. Still, Joy's record was not perfect. She had incorrectly identified one of the controls as a Parkinson's patient. The researchers wondered if the sample had been contaminated, or if Joy's nose had simply gotten tired. By Barran's recollection, Kunath's response was: "It's fine! It's one false positive!" Barran herself was slightly more cautious: Joy had mislabeled both halves of the man's T-shirt.

Of more immediate interest, though, was the question of what was causing the smell in the first place. The odor seemed to be concentrated not in the armpits, as the researchers had anticipated, but at the neckline. It took them several weeks to realize that it perhaps came from sebum, the lipid-rich substance secreted by the skin. Sebum is among the least studied biological substances. "It is actually another waste disposal for our system," Barran says. "But no one had ever thought that this was a bodily fluid we could use to find out about disease."

Barran set out to analyze the sebum of Parkinson's patients, hoping to identify the particular molecules responsible for the smell Joy detected: a chemical signature of the disease, one that could be detected by machine and could thus form the basis of a universal diagnostic test, a test that ultimately would not depend on Joy's or anyone else's nose. No one seemed to be interested in funding the work, though. There were no established protocols for working with sebum, and grant reviewers were unimpressed by the tiny pilot study. They also appeared to find the notion of studying a grandmother's unusual olfactory abilities to be faintly ridiculous. The response was effectively, "Oh, this isn't science— science is about measuring things in the blood," Barran says.

Barran turned to other projects. After nearly a year, however, at a Parkinson's event in Edinburgh, a familiar-looking man approached Kunath. He had served as one of the healthy controls in the pilot study. "You're going to have to put me in the other category," he said, according to Kunath. The man had recently been diagnosed with Parkinson's. Kunath was stunned. Joy's "misidentification" had not been an error, but rather an act of clairvoyance. She had diagnosed the man before medicine could do so.

Funding for a full study of Joy, the smell, and its chemical components now came through. "We saw something in the news, and we thought, Wow, we've got to act on that!" says Samantha Hutten, the director of translational research at the Michael J. Fox Foundation. "The NIH is not going to fund that. Who's going to fund it if not us?" At twenty-five NHS clinics in England and Scotland, Barran arranged for nurses to swab the upper backs of Parkinson's patients with small squares of medical gauze and to mail her the samples by standard post. (Unlike blood, sebum does not require a cold chain.)

In Manchester, she and her colleagues fed the samples into a gas chromatograph–mass spectrometer. A GC-MS machine separates a substance into its component molecular parts for identification. It cannot tell you how they smell, however, or even if they smell at all. Like all sense perceptions, smell is the brain's internal representation of the outside world; without a nose and brain to translate it into scent, a molecule is simply a molecule. (What it is about any particular molecule that causes it to smell the way it does is one of the great outstanding mysteries of the senses.) Barran added an odor port to the usual GC-MS setup, a rather goofy two-foot length of tube that resembled an elephant's trunk emerging from the side of the machine. Joy was positioned there, breathing in each type of molecule as it came off the separation column.

Of the more than two hundred molecular fragments the machine distinguished, Joy reported a strong Parkinson's-like scent in the presence of just three: eicosane and octadecanal, which are known to have weak waxy or oily smells, and hippuric acid, which is not typically reported to have any smell at all. Each of the chemicals was found in notably higher concentrations in the sebum of Parkinson's patients than in controls, the researchers wrote in their 2019 report. This was the source of the Parkinson's scent.

But why those particular chemicals were arriving on the skin, and what exactly they represented, remained uncertain. In a 2021 metabolomic analysis of Parkinsonian sebum, published in *Nature Communications*, Barran and her colleagues found evidence of changes in two important human metabolic pathways, the first implicating mitochondria, the second additionally implicating the organelles known as lysosomes. The pathways in question exist in cells throughout the body. But they are particularly active in our brains, where disruptions to mitochondrial and lysosomal function are known features of Parkinson's disease. The byproducts of these disrupted pathways—the molecules Joy detected—were, it seemed, being somehow transported to the surface of the body. With her nose, the researchers came to suspect, Joy was smelling the very death of the brain.

Joy has attained a certain prominence in the Parkinson's

realm. The work she inspired is, in the words of the research director at the Cure Parkinson's foundation, Simon Stott, "likely to become that stuff of legend." Joy is listed as a co-author on all the papers. She was named to the clinical-science subcommittee of the World Parkinson Congress; she has given the inevitable TEDx talk; she has the ear of some of the world's most respected scientists. "We need to trace it back from what she detects to its origin, to see what caused that," the Nobel laureate Randy Schekman, who serves as scientific director of the Parkinson's research initiative known as ASAP, told me. "She's a rare commodity." She has twice been flown to Tanzania for testing on tuberculosis. (In East Africa, the nonprofit organization Apopo uses giant pouched rats to diagnose the disease; Joy outsmells the rats.) Amazon has been in touch with her co-authors about the possibility of adding a smell functionality to its Alexa devices.

The fascination with Joy is of course attributable to the fact that she can smell the way she can, but it is heightened by the fact that no one knows why. She is hyperosmic by any reasonable definition, but hyperosmia has been the object of so little serious scientific investigation—the smell taboo in action, no doubt, at least in part—that it lacks even a set of agreed-upon definitional criteria. Its specific cause is unclear. Professor Thomas Hummel, of the Dresden University of Technology, a preeminent clinical investigator of olfactory function, told me his research suggests that self-reported hyperosmics have more connectivity in the higher brain regions responsible for olfaction. "These people pay more attention to odors; they take more out of the olfactory signal," Hummel said. But at this stage, this is merely a hypothesis. (Hyperosmics also seem to have more consensual, compromise-seeking personalities, he noted. "It's a very weak correlation," Hummel said, "but it's there." They also report more enjoyment of sex.)

Joy has enjoyed her fame, but the smell work also radicalized her, in its way, and she has a reputation for being a bit intransigent in her advocacy. The initial scientific skepticism toward her was of a piece, she thought, with what she already held to be the medical corps' hopeless wrongheadedness about Parkinson's disease. For Joy, as for many caregivers, the psychological aspects of the illness were by far the most difficult to manage, much less accept, and these happened to be precisely the symptoms neurologists seemed least interested in acknowledging, let alone addressing. "You're saying things to doctors and nurses, and they're not believing you," Joy told me. Eventually, Les's dementia and sexual compulsion intensified to such a point that she worried about leaving him alone with his young grandchildren. "Are neurology really willing to accept that they allow people to get that bad, and for the families to be going through that?" she asked. "It's unethical!" Les was hospitalized at one point, and Joy arranged for him to be monitored under a protocol for patients with psychological disturbances. A nurse had the protocol lifted "because 'he only had Parkinson's,'" Joy said. "They shut the door, put the cot sides up, and left him. And in the middle of the night, he walked across the bed, hit the wall with his head, and was on the floor for goodness knows how long."

To Joy's mind, still more proof of this medical obstinacy came from the discovery that she was not alone in her ability to smell Parkinson's disease. When the research first began to attract attention in the media, Barran and Kunath received messages from around the world from people reporting that they, too, had noticed a change in the smell of their loved ones with Parkinson's. The correspondents described an odor that was "musty," "oily," or "akin to sour milk." A man from New Mexico claimed to have diagnosed an acquaintance a year before his doctor. A Parkinson's patient wrote of the "social isolation" he experienced as a result of his own strong odor. One woman began, "I am a well-educated and sane person, so please don't disregard my email." But for the smell taboo, Joy thought, someone somewhere might have taken these people seriously, and the importance of the odor might have been realized decades sooner.

That Parkinson's has a distinctive odor and chemical signature has now been thoroughly demonstrated, but on its own this concept is not much more than a curiosity. The reason for the initial interest in Joy was not merely her claim that a smell existed but that it occurred very early in the course of the disease. But Joy's ability to smell Parkinson's had not yet been tested on early-stage patients: It is difficult to recruit such people into studies precisely because there is no good way to find them. They haven't yet been diagnosed with Parkinson's.

A few years ago, after reading about Joy, the neurologist Werner Poewe called Barran with what he thought might be a solution. Poewe, a prominent Parkinson's researcher and clinician at the Medical University of Innsbruck, had lately been interested in the condition known as isolated REM sleep behavior disorder, or iRBD, in which patients act out their dreams. Les had developed the condition years before his Parkinson's diagnosis, and iRBD is now known to be strongly predictive of Parkinson's. Nearly eight in every ten new iRBD patients will be diagnosed with Parkinson's or a related disease within a decade. (Reports of iRBD and related disorders increased during the Covid-19 pandemic, and some researchers fear a coming wave of Parkinson's diagnoses.)

In Austria, three dozen people were recruited, some with iRBD, some with Parkinson's, and controls with neither condition. A nurse swabbed the participants' upper backs, and the samples were sent to Barran and Joy for chemical and olfactory analysis. This was merely a pilot study, Poewe cautioned, but the results were "almost too good to believe." In subjects with iRBD, Joy found a scent that was similar to that of Parkinson's, but less offensive and clearly distinguishable. (She described it as "biscuity sweet.") The chemical signatures that appeared in GC-MS were likewise distinctive, and it was possible to distinguish Parkinsonian sebum from iRBD sebum from control sebum with perfect accuracy. Yet iRBD shared a great number of the chemical features of Parkinson's, at somewhat lower levels; the condition appeared, the researchers wrote in a draft of their findings posted last year to the preprint server bioRxiv, to be an "intermediary phenotype," a waypoint along the path toward the disease. The shared chemical features "could be early indicators of pre-manifest P.D.," they concluded, which is to say, an early diagnostic.

Much remains to be done. The results must be confirmed in larger cohorts and by separate research groups, and the actual predictive power of the test must be established: For a given level of these "early indicators," with how much certainty can it be said that a person will in fact develop Parkinson's? In what time frame? Answers to these questions will come only from longitudinal studies, and it will be years before the answers are known. But Barran, Kunath, and Joy have begun to imagine the possibility of using sebum tests to screen for Parkinson's on a broad scale. Kunath mentioned to me that he had recently turned fifty; the Scottish health system automatically mailed him a stool-sampling kit, to screen for bowel cancer. In a similar manner, it would be simple to send out cotton swabs for Parkinson's. Barran and the University of Manchester have spun off a company, SebOMIX, to commercialize such a test. (Barran is also enthusiastic about the prospect of detecting other conditions in sebum, which turns out to be a repository of all sorts of information. "To be honest, I really think that was the biggest discovery," she told me.)

Poewe is more circumspect. He envisions sebum analysis as perhaps one of a constellation of screening tests from which a "risk profile" could emerge, he says. "I don't think this will be *the* test." But if a sebum swab suggested Parkinson's in a patient with no outward signs, perhaps the patient would undergo more invasive testing. Perhaps she would then begin early treatment with whatever neuroprotective drugs were by then available, their development having itself been made possible by the ability to identify and study early-stage Parkinson's patients. And perhaps, as a consequence, the patient would never go on to develop the classical symptoms of Parkinson's at all. "I think that's realistic for the future—I think that will happen," Poewe says. "The question is when."

In the meantime, the question is what good it would do to tell an otherwise healthy person that he or she will someday fall victim to a terrible disease, and whether such a person would want, or should want, to know. "It helps research," Poewe says. "I'm not sure it helps people."

In the past half century, medicine has come to favor a high degree of transparency: The patient's autonomy is paramount, and he or she thus enjoys a "right to know," enshrined most famously in the doctrine of informed consent. At the same time, technological progress has vastly expanded the realm of the knowable. But knowledge can punish or paralyze. Too often, information has been viewed to be an "unmitigated blessing," the Oxford University law scholars Jonathan Herring and Charles Foster have written. "Life and medicine are not that simple."

I met Joy the spring before last, at her home in a tidy subdivision on Kinnoull Hill, in Perth, Scotland, overlooking the expansive green valley of the Tay. Les died in 2015, at sixty-five. Joy lives alone with a spoiled-fat sheepdog named Queenie, a short walk up from the house she and Les shared. She gave me a tour of her manicured garden and took out photographs of the family's old farmhouse in Cheshire, but she was most excited to show me her spice collection. "This is my delicious drawer," she told me in the kitchen, with an air of mischief. I had assumed that, given her extreme sensitivity, Joy would be attracted to only the mildest scents. "Totally the opposite!" she said, and she began pulling out jars of ground ginger, coriander, two distinct cinnamons, a cuminous garam masala. She opened a vial of allspice and banged on the side to release a bit of powder into the air. "It's all around us now!" she crowed. After a bit of rifling, she drew out some cardamom pods. She transferred a few to a mortar, pestled them, softly closed her eyes, and bent over the bowl in reverence.

I found Joy's ingenuous thrill to be strongly endearing, but I, like many others, was also a bit terrified of her nose. The radio journalist Alix Spiegel met Joy several years ago for a story on NPR. Alzheimer's, which Joy can detect, runs in Spiegel's family. "If she did smell it, would I be able to tell?" Spiegel wondered in her report. "How good was her poker face?" It is Joy's policy not to disclose disease odors to the people she meets, and she politely evaded Spiegel's questions. For whatever reason, she was more direct with me. One morning in her living room, she commented, unbidden, upon my "strong male scent."

I was aghast. "I wasn't going to bring this up," I said.

"No, no, it isn't like that," Joy assured me. "It's a normal male smell, almost like salt and a few chemicals. And it's sharp, but deep. It's when it gets to that creamy smell, and loses that sharpness, that I begin to think, Oh, what's wrong?"

It was relieving to receive a clean bill of health. (Given Joy's usual nondisclosure policy, I did wonder if she might be telling me a white lie, but I concluded, eventually, that she would not have offered one unprompted.) On the other hand, it was discomfiting to know that she had been smelling me at all. Our notions of privacy are calibrated to the sensory capabilities of the average other person. We learn to live with the reality that, if someone is just a foot away, he or she may be able to see the tiny pimple on our chin, or smell our breath, or perhaps hear the swash of our saliva. But we assume that at a slightly greater distance we are safe, that these intimate embarrassments will pass undetected. I am pleased to say that I am not a smelly person, or so I am told, but it was hard not to fret about what else, beyond my "male smell," might be accessible to Joy's nose. Nor is it always straightforward for Joy. She smells disease everywhere, without seeking it out: in the checkout at Marks & Spencer, on the street, on her friends and neighbors.

When we met, Joy informed me that Les's mother was not the only other member of the family to be diagnosed with Parkinson's. So too, she eventually discovered, were Les's maternal grandfather, his maternal uncle, his estranged younger brother. His was evidently a heredity form of the disease and, given its incidence in Les's family, almost certainly an autosomal dominant form, which is to say a form that would be quite likely to manifest in his children. In all probability, at least one of his and Joy's three sons would have inherited the gene.

Joy declined to discuss any genetic testing her sons may have undergone, and though she promised several times to put me in contact with them, she never did. I saw no dignified reason to press the matter any further. In the abstract, however, I can just as easily imagine them—fathers themselves—choosing to remain ignorant of their inheritance, and of their likely fate, as choosing to learn it. "Some of us like to feel the wind of providence in our faces, and others like everything planned," write the legal scholars Herring and Foster. "Each person should be allowed to choose how to approach his or her future." Joy, of course, will have no such choice. The wind of providence is always blowing; her nose cannot help making out whatever tragedies may float upon it. Whatever her own wishes, she will be made to know.

LEATH TONINO

This Is What It's Like to Camp in One of the Hottest Places on Earth

FROM *Outside*

LET ME ACKNOWLEDGE, right up front, that in this ghastly era of anthropogenic global warming I combusted a whole bunch of fossil fuel in order to descend from the cool green sanctuary of the Colorado Rockies, where I'm blessed to reside, and cross the hot, dry, fiercely sunburned interior West. My destination was the kiln of the Mojave Desert and, sequestered within that immensity of thirst, a line on the thermometer: 120 degrees Fahrenheit.

Was this a vacation? A gross display of privilege? According to the CDC, extreme heat waves cause approximately 1,220 deaths in the U.S. annually. Granted, I do not belong to the especially endangered demographic groups: infant, senior, unhoused, impoverished, employed outdoors. The list is tragic and long. But trust me, the trip wasn't idle amusement. I felt compelled to make raw somatic contact with our new and thoroughly dismaying climate regime, to face the faceless temperatures of the twenty-first century.

Sean is a social-studies teacher in Las Vegas who spends much of his summer break driving random dirt roads, exploring the desiccated, dust-choked hinterlands of Nevada and California. His style is the opposite of athletic, unless geography paired with existential contemplation constitutes a sport. He pokes around, parks the Hyundai, plants a parasol, eats and drinks, hikes a mile or three at dusk, counts shooting stars, sleeps, moves on. The very emptiness and quiet are his activity, the elemental place—overwhelming in a dozen different ways—his passion.

Chatting on the phone in early July, he informed me that the mercury in his apartment in North Vegas was registering 120 degrees, a record for the city. "A/C shut off yesterday," he said. "Kicked back on this morning. The grid . . . a surge . . . my unit . . . I dunno. In any case, I'm heading out for twenty-four hours." Air temps at Furnace Creek, in Death Valley National Park, were approaching the world's highest reliable measurement of 130 degrees, made there in 2021. "I bet it'll only be teens in the Mojave Preserve," he continued. "And single digits or lower at night."

This omission of the "hundred" prior to "teens" and "single digits" reminded me of how folks at Amundsen-Scott South Pole Station, where I once worked, eschew the phrase "below zero" because, quite simply, "above zero" doesn't occur in that part of Antarctica. I'd confronted (negative) 80 degrees during my stint on The Ice and handled it pretty well. In fact, I'd relished the challenge of strenuous labor, the steady, drudging effort that pumps blood to fingers and toes, lungs and brain. Our apocalyptic present is another matter. Strenuous labor is potentially lethal and the steady, drudging effort is that of patience: hunkering in the shade, trying your damnedest not to budge.

Sean isn't exactly a fan of the heat, but he accepts its authority, and this allows him to briefly sneak outside even when doing so is deemed reckless, or at least exceedingly unpleasant. We decided I should visit him ASAP to join one of his twenty-four-hour excursions into the reality that almost nobody is eager to embrace—call it our current and future home.

I wrote an email to my parents in Vermont after hanging up the phone, explaining the plan, tacking on a paragraph about anxiety and electrolytes. My dad replied: "Do be careful as we bubble at 108 degrees." I was unfamiliar with the verb "to bubble" in the context of human physiology, but caught his drift. My mom, whose hairdresser claims I am responsible for the grays she is paid to dye blond, cut to the chase with her usual no-nonsense wisdom: "You've never experienced that kind of heat. I don't think we are *meant* to experience that kind of heat. I'll just say this—show it the utmost respect."

I'm haunted by a section of Luis Alberto Urrea's *The Devil's Highway*, the story of fourteen migrants who perished along the U.S.-Mexico border, in which he describes the "schedule of doom": heat stress, heat fatigue, heat syncope, heat cramps, heat exhaustion, heat stroke, heat death. At one point Urrea writes, "If you're really lucky, someone might piss in your mouth." Likewise, I can't shake a *Vanity Fair* article by William Langewiesche about global warming and gnarly temps in the Sahara. He chose to travel to the Algerian city of Adrar during a heat wave and delivered a spooky-flat takeaway: "That was a mistake."

A decade ago, backpacking in the Grand Canyon with my partner Sophia, I glimpsed the beginning of the end. We had hidden in the mists of Thunder Falls through a brutal August afternoon and commenced our climb to the North Rim two hours before sundown. In the middle of the Redwall switchbacks her skin went purple and her arms went limp. The only rejuvenating shade was that cast by my thin frame. She curled in a ball beneath me. We waited for dark.

Sophia suffered a minor form of heat illness and revived. Heat stroke—a rise in body temperature beyond 104 degrees and a subsequent "collapse of basic biophysical functions," as Langewiesche puts it—is the true nightmare. The science is detailed and complicated, but the gist is that evaporative cooling, or sweating, eventually fails to counter internal heating. Symptoms may include confusion, aggression, slurred speech, rapid breathing, hallucinations, nausea, dizziness, fainting, seizure, and coma. Young, fit, vigorous people can and do succumb. Langewiesche again: "[T]here are no guarantees."

Sean and I agreed that aiming ourselves at 120 degrees was serious business. This agreement was unspoken, communicated by our methodical, albeit semi-frenzied, preparation. In his sweltering apartment at 6 a.m., we filled a burly plastic jug with seven gallons of water and loaded duffels with three tarps, five maps, and enough hats in enough varieties to start a haberdashery. We fixed sandwiches and a dinner of macaroni and cheese, thereby reducing the need to expend energy in the field. We chugged many consecutive glasses at the kitchen sink. We confirmed that the burly plastic jug wasn't leaking. We reconfirmed.

The parasol with a PVC stand (butt-sawed sharp for jamming into soft ground) was already stowed in Sean's car from his previous outing. He checked the tires, the full-size spare, the jack, the battery charger, and, neurotically, I checked the burly plastic jug. At half past seven, en route to buy Gatorades, salty snacks, fourteen pounds of ice, and a topped-off gas tank, the Hyundai's dash thermometer read 107. "You can't assume the vehicle will whisk you to safety," Sean said. "If it breaks, what next? You're walking, or hitching, or something. I'd guess most European and even American tourists in a rental overlook that contingency. Calling a tow truck without reception is tough."

It isn't just automobiles that cause trouble. On July 2, a private plane had an engine problem and made an emergency landing west of Salsberry Pass on California Highway 178, inside Death Valley National Park. The pilot and passengers were uninjured—rescue personnel promptly arrived, presumably bearing cold beverages—but others have not been so lucky. In the weeks following my trip with Sean, I frequented the park's newsfeed: a hiker dead, a hiker evacuated by helicopter, a hiker who received third-degree burns on the soles of his feet (sand dunes, flip-flops, agony), a motorist who drove off an embankment and then died of exposure. Unsurprisingly, soaring temperatures in the Grand Canyon have also taken a handful of lives this summer.

The sky was huge and hazy as we traversed the urban sprawl, huge and blue as we exited the Spring Mountains, dropped into Pahrump Valley, and steered toward California. Our target was vague, based on a hasty internet survey of projected highs. (Furnace Creek: 126 degrees.) In addition to heat, we sought solitude, remoteness, and a spread of anonymous dirt where it'd be difficult to believe in the existence of anything besides geology and convection-oven air. Sean mentioned a gleaming white playa—insisted it was the quintessence of *blistering*, its abiotic austerity unsurpassed—but ultimately we couldn't resist a twenty-eight-mile washboard road in southeastern Death Valley National Park, between the Greenwater Range and the Black Mountains.

At the turnoff, a yellow sign emblazoned with the silhouette of *Gopherus agassizii*, the threatened desert tortoise, greeted us instead of a ranger's ticket booth. Was an ancient reptile, for all intents and purposes a fifteen-million-year-old dinosaur, actually roaming this expanse of creosote scrub, subsisting on beavertail cactus flesh, going about her day ignorant of the weather alerts, the headlines, the untold human tragedies? For a moment, I felt the deep history of capital-H Heat, the scorch of the Mojave that was born at the close of the Pleistocene. We had the windows open. A shiver raced up my spine and bumped into the fat beads of sweat already rolling down.

The sweat kept coming, pouring from my armpits, pooling in my belly button, as we proceeded three miles to a tiny gravel drainage and pulled over at 9:30 a.m. It kept coming as we rigged a tarp system with parachute cord, trekking poles, tent stakes, the car's roof rack, and hot-to-the-touch rocks scavenged nearby. It kept coming as we paused and listened and heard a lone grasshopper's brittle clicking. It kept coming as we arranged furniture—ratty camp chairs, rickety table, the cooler serving as an ottoman—to create a surreal man cave.

Chores took less than forty-five minutes. We stripped to shorts, sat back, and peered out from our precious, precarious rectangle of shade. The rectangle morphed into a parallelogram. Sean scooched to the right. I scooched to the right. Gatorade the first segued to Gatorade the second. Soon I was coated in the finest grit, a glittering suit of nearly imperceptible particles carried by a nearly imperceptible breeze.

"Impressive that you do this solo," I said.

Sean wiped his brow. "What?"

"Impressive that . . ."

"No, *do* what?"

I nodded at nothing, everything, the dull intensity, the blaring silence, the weird sensation of being hemmed in by an invisible force, a gargantuan power, yet unable to engage it directly for fear of withering. "This," I repeated. "Do *this*."

Scribbles from my notebook . . .

10:45. At the South Pole, you're tethered to the station, the diesel generators and chocolate chip cookies, the imported warmth. Here, it's tarps and drinks, the microhabitat we've established. As a species, we survive by modification of environment and plasticity of behavior, period.

11:20. Head hurts. Hydration is impossible, hydration is mandatory. Square that circle. A quart an hour minimum? More? Diarrhea is a frightening prospect. You could literally shit yourself to death in the Mojave due to cheap tacos. Probably happens quick.

Noon. Gotta pee. Went shirtless and barefoot earlier and the result was a desperate sprint for cover. I'm dressed appropriately this time, popped collar, sneakers, etc. Off I go.

12:15. Pee was neither transparent nor sickeningly chartreuse. I'll consider that a win. Metaphors proliferate out there. A vise clamping the rib cage. A pizza stone pressed to the temple. A rough, sun-administered frisking. A claustrophobic hug from Satan—or maybe God.

1:10. Dumped half a cup of water into the gravel six inches from my seat. I'm also monitoring the apple Sean set on the table. Will it turn to fruit leather?

1:35. Officially gone. Damp patch? What damp patch? If half a cup of water spills in the desert but no one . . .

2:50. We've been discussing the question of who leads in the dance. In this instance, definitely the heat. That's tricky for the typical modern American. Usually, we're active agents, calling the shots: *I want to accomplish such-and-such task. I want to recreate in such-and-such fashion. See that hill? I'm gonna jog it! Right now! Giddyup!* Our will to push, to persevere, to achieve, is undeniably badass—we're talented in that regard—and therein lies the problem. Overconfidence. Better at giving commands than taking them.

4:40. Two minutes ago, reading a field guide aloud, the word "bush" came out as *buh-sh*, like saying "bus" with a lisp. WTF?

How do you pronounce it? I studied the page for a solid 30 seconds, stymied. Tried *boosh* and immediately knew it was wrong, which was reassuring. I'm fuzzy, sloppy. Decent enough, but far from normal.

5:05. Screw journaling. Language is too heavy a lift. Ditto this pencil in my mind (meant to write "hand"). And it's sizzling, the pencil. Same for the notebook, the legs of the chair, the car's fender and hubcaps, the Gatorade in my bottle, every surface everywhere. Sizzling. Screw it.

5:50. OK, finally ready for a little hike.

We meandered east, up a gentle grade: lizard tracks but zero lizards, parched soil crumbling underfoot. Initially, our hovel appeared as a curious anomaly, an arbitrary scuff on the clean sweep of the valley floor—then it shrank to an insignificant speck. Though the sun remained intimidating, a fist and a half above the horizon, the worst was over.

Moving through that desolate scenery—the craggy browns of the Black Mountains, the lumpy browns of the Greenwater Range, the innumerable beiges and ochres and umbers sloping away, away, away—I realized what had rendered writing preposterous. It wasn't merely heat. It was heat plus the dopey, slack-jawed vigil, eight straight hours scoured of the usual excitements and diversions: laptop, phone, music, bird singing from a tree, jet rumbling in the clouds, a toilet to flush, a doorknob to twist, an allegiance to something other than passivity. Heat may have sapped the *energy* required for meaning-making, but staring at the bleak, beautiful, radically non-linguistic landscape for seemingly longer than forever had sapped the *desire*. Thoughts, sentences? They diffused into the vastness.

Step by sluggish step, the change of pace and perspective returned us to language. At the base of a rubble-strewn ridge, we plopped down, gulped water, and riffed on the idea of heat as wilderness, a wide and rugged terrain that can't easily be escaped once entered, that leaves you small, cautious, humble—and if it doesn't, you pay the price. This idea led us to awe, that mix of terror and wonder often associated with nature's monolithic indifference and incomprehensibility. Obviously, it's unconscionable to celebrate the killer temps of the new climate. This weather murders soldiers in body armor, laborers in farm fields, panhandlers on the street, grandma and grandpa, the helpless, the trapped, whoever. So we didn't celebrate it. We simply recognized it as staggering, category-shattering, a phenomenon that deserves (thanks, Mom) the utmost respect.

Around 10 p.m. After stumbling back to the Hyundai, chowing hard on mac and cheese, grinning because we spied a bat, a fellow mammal, flitting against the orange sunset glow—I wished Sean a nice uncomfortable rest and wandered off to inflate my pad among the bushes. (*Bushes*, of course, pronounced like Busch Light, like George Bush, duh.) Lying there in my boxers with a cooler-dunked bandana pasted to my stomach, I watched the constellations spin. I've camped in the desert without a tent and I've camped in the desert without a mummy bag, but rarely have I camped in the desert without either, without the physical and psychological mediation they provide. The feeling was one of total vulnerability. Simultaneously, there was a quality of intimacy.

Technically, the day never ends and the sun never quits raining its life-generating, life-obliterating fire. Technically, what we call night, relief, is just the planet's prodigious shadow, a fleeting gift of shade, a very thick screen temporarily buffering the heat. The earth, I thought, in the drowsy-dazed manner of a guy teetering at the cusp of dreams—the earth is a tarp.

With that I fell asleep. Sort of. The bandana became a useless crust. The temperature dropped, but not by much, maybe high nineties in the run-up to dawn. Red ants bit my calves. Red ants bit my triceps. And a pain throbbed in my chest, an ache for tomorrow, when my best friend and I would hop in the air-conditioned car and the desert, framed by both the rearview mirror and the windshield, would continue to burn.

SARAH ZHANG

After the Miracle

FROM *The Atlantic*

THEY CALL IT the Purge.

You have experienced, in a modest way, something like it in the waning days of a bad cold, when your lungs finally expel their accumulated gunk. The rattle in your chest quiets. Your sinuses clear. You smell again: the animal sweetness of your children's hair, the metallic breeze stirring a late-summer night. Your body, which oozed and groaned under the yoke of illness, is now a perfectly humming machine. Living is easy—everything is easy. How wonderful it is to breathe, simply breathe.

Imagine, though, that you had never been able to simply breathe. Imagine that mucus—thick, copious, dark—had been accumulating since the moment you were born, thwarting air and trapping microbes to fester inside your lungs. That you spent an hour each day physically pounding the mucus out of your airways, but even then, your lung function would spiral only downward, in what amounted to a long, slow asphyxiation. This was what it once meant to be born with cystic fibrosis.

Then, in the fall of 2019, a new triple combination of drugs began making its way into the hands of people with the genetic disease. Trikafta corrects the misshapen protein that causes cystic fibrosis; this molecular tweak thins mucus in the lungs so it can be coughed up easily. In a matter of hours, patients who took it began to cough—and cough and cough and cough in what they later started calling the Purge. They hacked up at work, at home, in their car, in bed at night. It's not that they were sick; if anything, it was the opposite: they were becoming well. In the days that followed, their lungs were cleansed of a tarlike mucus, and the small tasks of daily life that had been so difficult became unthinkingly easy. They ran up the stairs. They ran after their kids. They ran 10Ks. They ran marathons.

Cystic fibrosis once all but guaranteed an early death. When the disease was first identified, in the 1930s, most babies born with CF died in infancy. The next decades were a grind of incremental medical progress: A child born with CF in the '50s could expect to live until age 5. In the '70s, age 10. In the early 2000s, age 35. With Trikafta came a quantum leap. Today, those who begin taking the drug in early adolescence, a recent study projected, can expect to survive to age 82.5—an essentially normal lifespan.

CF was one of the first diseases to be traced to a specific gene, and Trikafta is one of the first drugs designed for a specific, inherited mutation. It is not a cure, and it doesn't work for all patients. But a substantial majority of the forty thousand Americans with CF have now lived through a miracle—a thrilling but disorienting miracle. Where they once prepared for death, they now have to prepare for life. "It's like the opposite of a terminal diagnosis," Jenny Livingston told me.

Jenny spent her twenties in and out of the hospital for CF-related lung infections. During her frequent weeks-long stays, she made some of her best friends in the CF ward, only to watch them succumb, one by one, to the disease that she knew would eventually kill her too. More than anything, she hoped to live long enough to see her daughter graduate from high school.

Today, Jenny is thirty-six. Four years into taking Trikafta, she's the healthiest she's been in her adult life. Her daughter is fourteen, a lanky high-school freshman. They're both obsessed with Harry Styles, and after Jenny started on Trikafta, they flew together to see him live—twice. They learned to hunt deer with Jenny's partner, Randy. They often go up into the aspen- and fir-topped mountains that overlook their little town in central Utah. Jenny's last hospitalization—four years ago, just before she started Trikafta—is now more distant in time than her daughter's future graduation.

Having lived one life defined by cystic fibrosis, Jenny wonders: What is she going to do with her second life?

Jenny was born in 1987, the youngest of her parents' five children together and the third to have cystic fibrosis. Given the family history, the doctors knew to test her as an infant, wrapping her forearm in plastic until a sheen of sweat appeared on her skin: the classic "sweat test" for cystic fibrosis. The faulty protein in CF cannot control the balance of salt and water in the body, which results in mucus that is unusually thick and sweat that is unusually salty. In medieval Europe, centuries before anyone understood why, a proverb foretold the fate of children with salt on their skin: "Woe to the child who tastes salty from a kiss on the brow, for he is cursed and soon will die."

The 1980s, suffice it to say, were not the Middle Ages. By the time Jenny was born, her two older sisters with cystic fibrosis—Shannan, eight, and Teresa, seven—were on a strict schedule of mucus-clearing chest therapy and medications that had kept them alive past toddlerhood. Shannan wasn't diagnosed until she was thirteen months old. "I knew when she was born that there was something wrong," their mother, Lisa, told me. As a newborn, Shannan projectile vomited and blew out her diapers constantly. When she got older, she was often so insatiably hungry that she would cry when a spoon scraped the bottom of a near-empty food jar. She scarfed down five pancakes at a time. In the baby photos in Lisa's scrapbook, she is all skinny legs and big, swollen belly—a classic sign of malnutrition.

Shannan *was* starving, it turned out. Food was passing through her body undigested because her pancreas had been damaged as a result of thick mucus blocking the ducts that release digestive enzymes. Cystic fibrosis was originally named, in fact, for the fibrous cysts that a 1930s pathologist saw in the pancreases of babies who had died. An early epiphany helped doctors overcome the malfunctioning pancreas, though: The missing enzymes could be replaced with pills. By the time of Shannan's diagnosis, CF was known as a disease of the lungs, in which sticky mucus made fertile ground for bacteria, and the cycle of infection and scarring, infection and scarring would eventually cause the lungs to fail.

Lisa relayed the news of Shannan's diagnosis over the phone to her husband, Tom, who was at work. As she repeated the doctor's words, their awful meaning sank in. Their daughter would not live long. They would watch her die. In that moment, the two of them broke down on the phone, the physical distance between them collapsed by grief.

Shannan died when she was fourteen. "I remember the sound of her oxygen machine more than her voice," Jenny told me. The rumble and puff of the machine had run in the background of their home, punctuated by chronic coughs from all three girls with CF. But neither Teresa nor Jenny was ever as sick as Shannan was in childhood—due perhaps to chance or to being diagnosed and starting treatments earlier in life. Even when they were newborns, their mother coaxed applesauce sprinkled with enzymes into their mouth, so they could absorb nutrients from their milk.

Not long after Shannan died, Lisa and Tom divorced—their marriage had been strained even before the loss of their daughter—and they both eventually remarried. Despite the upheavals in her family, Jenny remembers her childhood as quite normal. Yes, she had to take the enzymes with every meal, and she had to clear her lungs of mucus every day—first by having her parents pound on her chest and back and later by using an oscillating vest that shook her body. As inhaled CF drugs were developed, they were added to her daily regimen. She went to the hospital for annual preventive "tune-ups," but she was never sick enough to need emergency hospitalizations, and CF did not seem to hold her back.

Lisa thinks of Jenny as her sassy daughter. Her youngest was always stubborn, always a go-getter. Through the Make-A-Wish Foundation, she was able to get a horse, which she entered in local shows and rode through the foothills just outside town. In the summer, the salt from the dried sweat on her arms became crystals that glimmered in the sun, a subtle reminder of the disease still inside her. The invincibility of youth, however, made her think she had perhaps escaped her oldest sister's fate.

At nineteen, Jenny married a local boy she had fallen in love with, and at twenty-one, she was shocked to find herself pregnant: "A very, very happy surprise." She had always longed to be a mother. As a young girl, she once drew a picture proclaiming that she would grow up to have six children. The drawing "broke my heart," says her stepmother, Candy. Even if Jenny lived long enough, cystic fibrosis often causes fertility issues—in many women, thickened cervical mucus is thought to prevent pregnancy, and in almost all men, sperm ducts never develop because of blockages that occur in utero. And at the time, doctors often recommended against pregnancy for health reasons.

But Jenny pushed the worries out of her mind. She was simply happy. She set up a crib and painted the nursery. In retrospect, the fevers and shortness of breath she began to feel were not just the normal discomforts of pregnancy, but she didn't clock it then. She had an uneventful labor, and gave birth to a healthy baby girl. They named her Morgan.

The trouble started in the following months. Six weeks after giving birth, Jenny went back to work. Between nursing and soothing and diapering a newborn, she could no longer keep up her treatment routine. She sometimes also skipped medications when she couldn't afford them with the pay from her job as a bank teller and her husband's as a welder.

Then she caught a bug. It was 2009, the year of swine flu, so it could have been that or a more mundane cold, but either way, it triggered something deep in her lungs. She started feeling short of breath. By the time she got to a CF specialist at a hospital two hours away, in Salt Lake City, she could not walk from the car to the front door. She was too weak to stand for her lung-function test. She collapsed into her hospital bed, and for the next several days, she was unable to use the toilet or shower on her own. Convinced that she would die a hundred miles from her three-month-old daughter, she had a terrible revelation: "This is why they said, 'Don't have kids.'"

This was Jenny's first CF pulmonary exacerbation, when lung

function plummets from an acute infection. Doctors inserted her first PICC line, a catheter that runs from the upper arm to the heart, delivers antibiotics, and stays in place longer than an IV. She recovered, but just months later, she was back in the hospital with another exacerbation. Then another and another, and on this went for the next several years. Jenny counted for me the PICC-line scars still visible as white dots on each arm—at least ten on the left, sixteen on the right. When the veins in her arms started to reject PICC lines, doctors placed a port under her right collarbone for easy access to her central vein.

Each infection scarred her lungs; each exacerbation eroded her lung function. The disease that had been a minor plot point in her life became one of its major storylines, and the people in the hospital became recurring characters. At the University of Utah's CF center, she met Warren, one of her best friends, whom she came to know so well, she could identify his cough through the hospital walls. He was "so dang funny," Jenny said, unafraid of joking about the death that would befall them both. Where she was a rule follower, he was a troublemaker. Once, he commandeered a hospital floor scrubber, waving at patients in their rooms as he drove past. Another time, he managed to procure a bootleg copy of *The Avengers*. Stuck in the hospital over the film's opening weekend, he and the other CF patients organized a movie night. James brought his Xbox to play the bootleg DVD. Heather ("the biggest Swiftie") and Angie ("gorgeous, tall blonde") joined too. They found a waiting room with a TV, and the nurses passed around microwave popcorn.

Jenny and her friends made sure to sit several feet apart. People with cystic fibrosis have had to practice social distancing since long before Covid, because they are considered a danger to one another. Their lungs harbor destructive and often antibiotic-resistant bacteria that can become impossible to uproot once established. Certain names are spoken with an air of doom: *Burkholderia cepacia*, *Pseudomonas aeruginosa*. When doctors in the 1990s realized that people with CF were infecting and killing one another by simply gathering, they stopped allowing patients to go within several feet of one another unmasked. Camps for children with cystic fibrosis, which Jenny still remembers fondly, were all shut down. In the hospital, she once again found a community in the disease that was taking over her life. But many of those friendships ended too soon: Of the five people at the *Avengers* movie night, Jenny is the only one alive today. Warren, James, Heather, and Angie have all died.

As Jenny struggled with her health, the new reality of chronic illness took a toll on her marriage. She and her husband eventually divorced. After a particularly harrowing hospitalization in 2012, her doctors encouraged her to stop working and go on disability. Something in her life had to give, they told her, or it would be her body. Her disease and her daughter became her whole world.

Even as a young child, Morgan could sense when her mom was heading toward another exacerbation. If she noticed that Jenny was more tired than usual or coughing more than usual, she began to dread their coming separation. When she was three years old, she asked, "Do all mommies live in the hospital sometimes?" When she was six, after Warren's death, she asked, "Can you die from CF?" She understood that their existence together was fragile.

Jenny answered truthfully: Yes. But she assured her daughter that she was taking care of herself as best she could. Still, she made plans for what was probably inevitable. If she died, her daughter would live with her aunt and uncle. If she died, she wanted a funeral just like Warren's, with music, candy, and an open mic for everyone to share their favorite memories.

A cure for cystic fibrosis had supposedly been imminent since 1989, when Jenny turned two. That year, scientists identified the recessive gene behind cystic fibrosis, which encodes a protein called CFTR that controls the flow of salt and water. The discovery seemed so explosive that a Reuters reporter rushed to publish the scoop more than two weeks before the scientific papers were due to come out; two press conferences followed.

In the decades after, however, researchers came to understand the wide gulf between identifying a genetic problem and knowing how to solve it. Early attempts in the '90s at using gene therapy to fix mutations failed again and again, both for CF and for other genetic conditions that once seemed tantalizingly close to a cure.

Then, CF researchers changed tack: Instead of correcting the gene, why not correct the mutated protein itself with small fixer molecules? This had never been done before—with any disease— but the nonprofit Cystic Fibrosis Foundation deemed the strategy promising enough to strike an unusual venture-philanthropy agreement with a company that would attempt it, which was eventually bought by Vertex Pharmaceuticals. The foundation funded the research in return for a share of the revenue.

The move paid off. In 2012, Vertex released a drug called Kalydeco that worked stunningly well—improving lung function and erasing many symptoms in the small group of CF patients who could take it. That was the catch: The FDA approved Kalydeco only for the roughly 4 percent of people with CF who carried a rare and specific mutation. Still, it provided a jolt of optimism. Kalydeco was the first drug ever tailored to a person's inherited genetic mutation, and the breakthrough portended a new age of "personalized medicine." It also inspired other patient-advocacy groups to copy the venture-philanthropy model. In 2014, the Cystic Fibrosis Foundation sold the rights to royalties from Kalydeco and future Vertex CF drugs for $3.3 billion, which it could invest in new research.

After Kalydeco, the next CF mutation to target was obvious. About 1,700 unique mutations have been found in people with CF, but some 90 percent of patients—including Jenny—carry at least one copy of a mutation, known as F508del, that leaves their protein channels too seriously distorted for Kalydeco alone to correct. Fixing this shape would be a much bigger task. In 2013, Jenny joined the clinical trial for a two-drug combination from Vertex, made up of Kalydeco plus a second fixer molecule. It failed to especially improve her symptoms, though it did work enough to stabilize her falling lung function. "It seemed to push pause," she said. She stopped getting sicker, but she was still sick. The research went on.

A few years later, word began spreading of a forthcoming three-drug combination from Vertex. In clinical trials, neither patients nor doctors are told who is on the placebo and who is on the experimental drug. But in this trial, everyone could tell. The triple combo made patients' lung function jump by a shocking 10 percentage points. Overnight, they woke up smelling for the first time the distinctive scent of their home. They could even taste their sweat becoming less salty. This was Trikafta.

In the fall of 2019, Trikafta was approved by the FDA just ten days before a large annual gathering of CF experts in Nashville. Doctors who attended told me the atmosphere was electric. Jenny happened to be there to speak on an unrelated panel, and she remembers seeing the geneticist Francis Collins walk onstage with a guitar. Collins is best known as the longtime director of the National Institutes of Health, where he oversaw the sequencing of the human genome in the '90s (he has since retired from the NIH). But he had made his name in 1989 as one of the scientists who discovered the gene for cystic fibrosis.

In those long years when progress was halting, Collins, who is also an amateur musician, wrote a song to inspire a gathering of CF researchers. He sang "Dare to Dream" again that day in Nashville, his baritone raspier with age. When he got to the verse that he had rewritten for this occasion—"That triple treatment has taken 30 years"—cheers broke out in the convention center. In the crowd were people who had waited their whole career, even their whole life, for this moment. *We dare to dream, dare to dream.* As they swayed to the music, perhaps no one quite understood the magnitude and velocity of the change to come.

Jenny received her first box of Trikafta on November 17, 2019, at the end of yet another two-week hospital stay. She had gotten sick again in Nashville. Actually, she had been fighting off a cold before she left, and despite assiduously staying in her hotel room to keep up her treatment routine, she felt an infection settling into her lungs. At the conference, she heard a lot about Trikafta, but she didn't expect to get it so quickly. CF centers were being inundated with calls from patients asking for the new drug.

In the hospital in Utah, she recorded a video that she sent to her sister with CF, Teresa, who now lived in Ohio. She is sitting on her hospital bed. "My Trikafta is here," she says, her voice shaking and her eyes tearing up. The miracle drug she had been promised her whole life was now in her hands.

Teresa was also able to start the drug not long after. For her, Trikafta's impact was immediate and unmistakable. The Purge started on the drive back from the doctor's visit where she took the first dose. The mucus coming up was so thin that she was confused; it was nothing like the sticky gunk she'd had to work so hard to cough up. A month later, she went back for a sweat test, and her salt level was normal. Based on the results, you would not know she had cystic fibrosis.

"I think of it like, 'Oh, back when I used to have CF,'" Teresa said on a recent call with Jenny and me. "I don't feel like I have CF. I feel completely normal." She has been able to stop using her vest and inhaled medications, freeing up that time for her adopted children and the farm where she lives with her family. Before Trikafta, every small exertion was a negotiation with her lungs. Should she go upstairs? How many breaths would that take? Now she's running around milking the goats, trimming their hooves, throwing thirty bales of hay into the barn.

On that same call, the sisters got to talking about an upcoming trip to see their grandmother, and Teresa asked Jenny a question that would have been inconceivable before Trikafta: Could they stay in the same hotel room? To avoid infecting each other with the bacteria in their lungs, the two had not shared a room since Teresa left Utah fifteen years earlier. At family gatherings, they kept their distance. They didn't even touch the same serving utensils, sending their partners to get their food. Now, Jenny told her sister, "I would totally stay in the same hotel room."

When Jenny started Trikafta, it took her longer than it took Teresa to notice much change. She didn't have the dramatic capital-P Purge because, she thinks, the hospitalization had already temporarily cleared her lungs. But two months after she started the drug, when a snowstorm blanketed their town, her family drove out to their favorite sledding hill. Jenny had never liked sledding; she would stand in the cold while everyone else ran around having fun, their easy breaths turning into white puffs in the air. This time, her nephew called out and she jogged over.

It wasn't until she got to him that she realized she had jogged *up*—all the way to the top of the hill. "I don't run, and I don't climb hills. And I just ran up a hill and felt super fine," she says in a video she took right after. "I'm going to see if I can do it again. Ready?"

"Yes," her daughter, Morgan, answers next to her. They take off. "Mom!" Morgan shouts a few seconds later, as the distance between them grows larger. "You're beating me, Mom!" At the top of the hill, Jenny looks back to see Morgan still catching up. Jenny went down the hill and ran back up again, simply to prove that she could. "At one point, I just plopped up here on my bum and cried," she told me during my visit in October, pointing to the spot on the hill where it had all hit her. In front of us, big gray mountains jutted into the blue sky. The sledding hill, she admitted, did not look that impressive. But for all of Morgan's life, Jenny had been on the sidelines. She'd watch as Morgan swam in the lake or rode her bike, her low-grade fever making her too tired to join. That day on the hill, they finally ran together.

From there, Jenny began noticing changes in her body, big and small. The tips of her fingers, which had always been slightly swollen and round—a sign of low oxygen—thinned out as her lungs improved. She didn't need as many enzyme pills to digest her meals. Her chronic cough disappeared. She hadn't realized how much she had always suppressed her laughter to avoid triggering her cough. Now she can laugh—big belly laughs that match the warmth of her personality. "Oh my gosh, my laugh drives her crazy," she told me in the car, laughing, after picking up Morgan from school. "That's because you laugh at stuff that's not funny," her daughter shot back. Jenny laughed again.

Trikafta had effects that even doctors did not anticipate. In the months after the drugs became widely available, some patients unexpectedly got pregnant; the drug that thins lung mucus, it turns out, also thins cervical mucus. Then, patients started *trying* to get pregnant. The drug made many people with CF feel so healthy that they no longer worried about the physical toll of pregnancy and parenthood or the agony of leaving behind young children. Doctors began speaking of a Trikafta baby boom.

Doors opened to other once-impossible futures. A twenty-two-year-old told me he decided to train as an aircraft mechanic, a job that would have been far too physically demanding when he was being hospitalized multiple times a year. One woman started dating. "I don't want to fall in love with somebody, knowing that I'm not going to be around very long," she had thought. Now she and her boyfriend have been together for four years. A father who was being evaluated for a lung transplant before Trikafta felt healthy enough to spend the summer of 2020 tearing down and rebuilding his family's deck, and now expects his CF lungs to see him through graduations and grandkids.

Trikafta is a lifelong medication, and it is not meant to undo organ damage that has already occurred. But the earlier treatment begins, the healthier one stays. A handful of pregnant women have now used Trikafta to treat their unborn children with cystic fibrosis. Last fall, I corresponded with one such expecting mother, who does not have CF but whose son was diagnosed by genetic testing. She started Trikafta at twenty-six weeks. When her son was born in October, his lungs and pancreas were perfectly healthy.

Officially, Trikafta is approved in the U.S. for patients as young as two. Unofficially, some parents give their newborns Trikafta, either indirectly through breast milk or directly by grinding up the pills into tiny doses. So long as they stay on the medication, these children may never experience any of the physical ravages of the disease. Recently, Make-A-Wish announced that children with CF would no longer automatically be eligible for the program, because "life-changing advances" had radically improved the outlook for them.

CF centers these days are unusually quiet. Fewer patients need once-routine weeks-long hospitalizations. Instead of thinking about lung function, more and more are worrying about the maladies that come with middle and old age—colon cancer, high cholesterol, heart disease. Obesity has been a confounding new issue. Before Trikafta, patients were usually underweight, and they were told to cram as many calories in as possible, by whatever means possible. Every additional pound was a small victory. One patient described microwaving pints of Ben & Jerry's to drink mixed with heavy cream; when even that failed to make her gain weight, she got a feeding tube. Now people on Trikafta worry about getting too many calories.

In February, Vertex announced the results of a clinical trial for a next-generation triple-combination therapy, which may be even more effective than Trikafta. All of these changes have made for an existential moment for doctors too: The disease they were trained to treat is no longer the disease most of their patients have.

Doctors told me they could think of only one other comparable breakthrough in recent memory: the arrival of powerful HIV drugs in the 1990s. Like Trikafta, those drugs were not a cure, but they transformed AIDS from a terminal illness into a manageable chronic one. Young men got up from their deathbeds, newly strong and hale. AIDS hospices emptied—and then went bankrupt.

This was a remarkable turn of events. But it elicited a complicated mix of emotions, not all of them joyful. Some patients who were no longer dying grew depressed, anxious, and even suicidal at the thought of living. This phenomenon became known as "Lazarus syndrome."

Death is an end, after all. Life comes with problems: Patients who spent lavishly during what were supposed to be their last days now had no money to live on. Those who stayed with a lover in sickness found that they could not actually stand them in health. They fretted about insurance and paperwork and chores, everyday annoyances that would no longer be obliterated by imminent death. In 1996, the writer Andrew Sullivan, who is HIV-positive, described life after the advent of the HIV drugs in his essay "When Plagues End":

> When you have spent several years girding yourself for the possibility of death, it is not so easy to gird yourself instead for the possibility of life. What you expect to greet with the euphoria of victory comes instead like the slow withdrawal of an excuse. And you resist it. The intensity with which you had learned to approach each day turns into a banality, a banality that refuses to understand or even appreciate the experience you have just gone through.

For some HIV patients, their reversal of fortune seemed unreal. "He doesn't trust what's happening to him," one doctor said about a patient who had made a dramatic recovery, yet found himself in psychological distress.

Doubts like these crept into the minds of many people on Trikafta too. What if the new drug stopped working? Or had horrible side effects? Or stopped being covered by insurance? Trikafta's sticker price is more than $300,000 a year. Insurance typically covers most of that cost—minus what can be significant co-pays and deductibles—and Vertex offers co-pay assistance. But patients' lives ultimately depend on decisions made by nameless bureaucrats in rooms far away: insurance plans can suddenly change what they cover, and in 2022, Vertex announced that it would substantially reduce its financial assistance.

A forty-three-year-old woman I interviewed asked not to be named, because she feared that speaking about her improved health would cause her to lose disability benefits, which would also get her kicked off the government insurance that pays for Trikafta. Her health has not improved as dramatically as others' has, and she still has frequent infections and occasional bleeding in her lungs. If she returns to work but her health declines, it could take a long time to get back on disability—time she would have to go without Trikafta. She would also need a job with health insurance good enough to cover the expensive drug—but could she even get one as a fortysomething with no recent employment history?

For other patients, new health granted new independence, which could be scary too. As a child, Patrick Allen Brown was sick enough to miss long stretches of school. His parents didn't expect him to do chores, let alone support himself with a job one day. So much of his life was spent in the hospital that movies became his way of understanding the outside world. In his teens and twenties, he drank heavily.

After Trikafta restored Brown's physical health, he was no longer a chronically ill adult who lived with his parents. He was a pretty healthy adult who still lived with his parents. He was thirty-two, and hadn't finished college. Now he had to budget, commit to a career. He decided to get sober. When one of his parents needed back surgery recently, their roles flipped: He became the caretaker. Brown has now graduated from culinary school and found work as a chef, but he feels as if he is still catching up to his peers.

The great blossoming of possibilities on Trikafta also dredged up regret about decisions too late to undo. Kara Hansen, forty-one, has a daughter who was adopted, and she had always wanted another child. But in 2016, she had to be repeatedly hospitalized: in April, then again in May, July, and August. She gave up on having a second child—how could she, if she couldn't even guarantee living for the daughter she already had? Then, in 2018, she joined the original trial for Trikafta, becoming one of the first people in the world to experience its miraculous effects. If she had known her health would improve so dramatically and hold steady six years on, she would have tried to get pregnant, but she feels like it's too late now. To plan for such a miracle would have been foolish, but to live in its unexpected aftermath can still be painful.

After a year on Trikafta, Jenny told Teresa something that she acknowledged sounded "insane" but that her sister understood immediately: "To no longer be actively dying kind of sucks," she said. The certainty of dying young, she realized, had been a security blanket. She'd never worried about retirement, menopause, or the loneliness of outliving a parent or a partner.

Cystic fibrosis had defined her adult life. Now what? For so long, she'd just been trying to see her daughter graduate from high school. Now she faced seeing Morgan go off and live her own life. What then? Jenny had become active in patient advocacy, and soon after the start of the pandemic, she volunteered to moderate an online patient forum on mental health for her CF center in Utah. It went so well that her longtime social worker at the center felt compelled to give some career advice: try social work.

Jenny enrolled in an online master's program in 2022, and this past fall she chose a practicum with a hospice agency. Having watched the death of so many friends and contemplated her own, she felt prepared to shepherd people through the sadness and awkwardness and even humor that accompany the end of life. She understood, too, the small dignities that mean the world when your body is no longer up to the task of living. One hospice patient, she noticed, often had trouble understanding conversations because his hearing aids were never charged correctly. She got the situation fixed, and on a recent visit, he wanted to listen to music, playing for her the favorite songs of his youth. On another man's shelf, she recognized a birding book, and she made plans for a window feeder to bring birds to him.

Jenny doesn't share the details of her life with patients, but in their experiences with death, she has seen her own refracted. One hospice patient, a devout elderly woman, was estranged from her adult son, who no longer believed. Jenny herself grew up religious—Mormon, in her case—but she is not anymore. Her family is still Mormon, as is virtually everyone in the town she has lived in since childhood, which has 3,500 people, several Mormon churches, and a Mormon temple. She is liberal, whereas most of her relatives voted for Donald Trump.

Still, Jenny has made a point of staying close to her large, tight-knit family. Knowing she would die young had long ago clarified that she wanted to leave with no regrets, no grudges, and no words left unsaid to the people she loved. In the foothills outside town one day, she pointed in the direction of her house, her brother's house, her mom's house, her dad and stepmom's house, all minutes away from one another.

Although Trikafta looks to be a very safe drug for most people, it does have side effects. It can cause cataracts as well as liver injury. More perplexing, Trikafta may affect the brain.

For Jenny, starting Trikafta coincided with a wave of intense insomnia, brain fog, and anxiety. For months, she could sleep only two or three hours a night. She'd lose her phone and find it in the freezer. Her lungs were so much healthier, but her brain was going haywire. Soon, she realized that other CF patients had begun sharing stories online of depression, anger, or suicidal thoughts that emerged at the same time they started taking Trikafta.

Doctors sometimes chalked up these symptoms to the existential unease of no longer dying, or the fear and isolation everyone felt in the early days of the pandemic. But Jenny's doctor took the side effects she reported seriously enough to suggest that she halve her Trikafta dose, and soon after, they subsided. (Some of her CF symptoms did return, but they were muted enough that she could pare down her regimen of treatments.)

The link between Trikafta and these symptoms in the brain is still not fully proven or understood. "We've done an in-depth analysis of the preclinical data, clinical data, and real-world-evidence data, and we don't find any causal relationship," Fred Van Goor, a vice president and the head of CF research at Vertex, told me in January. And an analysis co-authored by the company's scientists last year found similar rates of depression and suicidality in CF patients with or without Trikafta. But in November, a group of scientists published a review arguing that the possible neuropsychiatric effects of Trikafta deserved a "serious research effort." The protein behind CF is found in cells throughout the body, including the brain. Trikafta could be acting on the brain directly, the authors hypothesized, or it could be acting indirectly via changes to inflammation throughout the body or specifically in the gut. The drug may affect different subsets of patients differently, says Anna Georgiopoulos, a psychiatrist at Massachusetts General Hospital who co-authored the review. She believes that neuropsychiatric side effects afflict only a "small minority" of people on Trikafta, but says that studies are needed to know exactly how many.

In the meantime, some patients have quit Trikafta altogether, their neuropsychiatric symptoms too debilitating even on a lower dose. "Physically I was feeling the best I've ever felt," says Aimee Lecointre of her time on the drug, but mentally, "I felt on the verge of a panic attack almost every day." The contradiction confused her: How could she be so anxious and depressed when her health was getting so much better? When she finally decided to try stopping Trikafta, the nervous energy that had filled her body all day long dissipated. But her CF symptoms came back. During our phone conversation, she paused every few minutes to cough.

She and Jenny have known each other for years, going back to their mutual hospitalizations. The three of us were supposed to meet over apple-cider floats when I was in Utah, but Lecointre had health issues come up at the last minute, the kind of disruption that happens all the time for people with a chronic illness. For a while, her Instagram feed filled with people on Trikafta whose lives were transforming while hers stayed the same; she had to delete social media from her phone. She still feels sad, sometimes, that Trikafta didn't work out for her. But she was able to go back to one of Vertex's two-drug combos, and although it is less effective than Trikafta, she feels so much better. There is more to cope with, but the coping is easier.

For another group of CF patients, Trikafta simply does not work. About 10 percent lack the F508del mutation that the triple combination was specifically designed to fix. Over time, though, scientists have found that some less common mutations are similar enough to F508del that those who carry them still benefit from Trikafta. And in late 2020, word got out that the FDA would soon approve the drug for additional mutations.

Gina Ruiz remembers waiting and waiting for the list of new mutations that fall. She had spent the past year watching her peers on Trikafta be handed what she thought of as a "reverse Uno card"—reverse weight loss, reverse lung decline, reverse CF—while her own health continued to worsen. She was sitting in a car when she saw the list, and she scrolled through the 177 new mutations hoping to find hers. She was crushed when she did not. Ruiz and most people in the 10 percent have mutations that leave their CFTR protein too garbled or incomplete to correct with any combination of fixer molecules. Treating these mutations will require a different strategy altogether.

The Cystic Fibrosis Foundation continues to fund research into a cure for all, and scientists, including those at Vertex, are once again exploring genetic therapies, applying the lessons of past failures. But a genetic-therapy breakthrough specific to CF is still years, if not decades, away. After Vertex created that first drug for the 4 percent, the path toward Trikafta was clear. After Trikafta, terra incognita.

Ruiz is wary of getting her hopes up again. At age twenty-nine, she can no longer work. She lives with her parents. Her lung function has fallen to 30 percent. And in December, her weight reached a new low of eighty-nine pounds. "I went to Target last night and I was beyond exhausted," she told me the following month. Her knees hurt too, another complication of CF. As she's watched her peers on Trikafta get married and chase after toddlers, her own world has shrunk. Halfway through the store, she got so tired that she had to rest in a chair in the home-goods section before she could go on.

Other patients with rare mutations told me the CF communities they once relied on for support have become quiet, as the 90 percent have gotten on with their lives. "It's extremely isolating," says Steph Hansen, who was steeling herself for another hospitalization when we spoke in January. She describes it as a one-two punch: Her health is no better, yet she has lost the community that once buoyed her. She's connected with a handful of other patients who can't take Trikafta, but CF is already a rare disease, and they are the rarest of the rare.

The F508del mutation is most common in people of European ancestry, so people with mutations ineligible for Trikafta in the U.S. are disproportionately Black or Latino. Globally, the proportion of people ineligible is higher in Latin America, Asia, and Africa, where diagnosis and treatment for CF also lag. In most developing countries, even eligible patients cannot get Trikafta—because Vertex currently does not sell its expensive drug outside a few dozen countries, concentrated in Europe and the English-speaking world. (Vertex says it has a pilot program that "provides Trikafta at no cost to people with CF in certain lower income countries.") Its patents also block other companies from making a cheaper generic version. In early 2023, activists asked four countries to revoke or suspend patents for Trikafta in a coordinated campaign. One of the countries was India, where *The New York Times* wrote about a father named Seshagiri Buddana. His son would have been able to take Trikafta if he lived in the U.S., but he died in December 2022 one day before he would have turned nine.

All of this weighs on Jenny. What makes her different from those who have died, other than the luck of being born at the right time, in the right place, with the right mutations?

*

Two days after my visit to Utah, Jenny's father, Tom, had a heart attack while chopping firewood. He felt short of breath, and a trip to the hospital revealed that his major arteries were 90-percent blocked.

When Jenny texted me the news, she said she had been replaying our recent conversations about life and death. She was glad to feel, upon learning her father might die, that nothing between the two of them was left unsaid or unresolved. I thought of what Tom had told me in his living room. Before we had gone over to his house that day, Jenny had warned me that her dad was a jokester, not a man prone to earnest reflection. But when the conversation shifted to the impact of Trikafta, he turned to me, completely serious. "I was going to bury my kids. And I'm not. They get to bury me, which is the way it's supposed to be."

We all fell silent for a moment, as we felt the weight he had been carrying all those years. After burying his eldest daughter at fourteen, Tom could no longer watch movies in which children die. In Jenny's years of sickness, he had often driven her two hours to the hospital in Salt Lake City, but he rarely set foot inside. Hospitals are places where people go to be born or to die, he'd say, and all my children have already been born.

After his heart attack, Tom needed an emergency quintuple-bypass surgery. He did well, and came home to recover. He spent the time rethinking his priorities. Just before falling ill, he had skipped a family outing to an amusement park to work. Now he regretted it. He's become more open about his emotions; still a jokester, he's taken to saying that his heart has been opened in more ways than one since the surgery.

It's interesting, Jenny says. Her father has lived a longer and very different life from her own, but she recognizes what he is going through. People die from this, he started saying. I could have died from this. He got close enough to see death's shadow, only to be pulled back to a life whose familiarity suddenly felt unfamiliar. What would he do with his unexpected life? "Hey," Jenny told her dad. "I get it."

Contributors' Notes

MICHAEL ADNO is a writer from south Florida. He's written for *The New York Times* about the concentration of psychics in his hometown, for *The New Yorker* about how Florida's flamingos disappeared a century ago before returning in a hurricane, and for *The Bitter Southerner*, where his profile of folklorist Ernest Mickler won a James Beard Award.

ROSS ANDERSEN is a staff writer at *The Atlantic*. He was previously the magazine's deputy editor. As a writer for the magazine, he has reported in Russia, China, India, Pakistan, and Greenland. He is also the author of *The Long Search*.

ROBIN BERJON is a technologist specializing in the governance of technology. His work focuses on building durable democratic systems so that our digital sphere starts operating in the public interest at the planetary scale.

KATIE ENGELHART is a contributing writer for *The New York Times Magazine*, based in Toronto. Her work focuses on the ethics and philosophy of medicine. She was the recipient of the 2024 Pulitzer Prize for Feature Writing.

DR. MARIA FARRELL is an Irish keynote speaker and writer of fiction and nonfiction. She has worked in technology policy for twenty years, including at The World Bank; ICANN; the International Chamber of Commerce, Paris; and The Law Society of England and Wales. She has written for *Noema, The Guardian, Conversationalist, The New European, Slate, Medium, The Irish Times,* and *Irish Independent,* and has appeared as a tech expert on BBC, Sky News, NBC, and ABC. Maria is now working on her forthcoming book, *Rewild and Resist: How Ecology Can Tame Big Tech to Create a World We Want to Live In.*

RIVKA GALCHEN is a staff writer at *The New Yorker*.

REBECCA GIGGS is the author of *Fathoms: The World in the Whale*, and is currently at work on a book about twenty-first-century pets.

BEN GOLDFARB is an environmental journalist whose work has appeared in *National Geographic*, *The Atlantic*, *The New Yorker*, *Smithsonian*, and many other publications. His most recent book, *Crossings: How Road Ecology Is Shaping the Future of Our Planet*, was named one of the best books of 2023 by *The New York Times*, and received the Rachel Carson Award for Excellence in Environmental Writing and the Banff Book Competition's Grand Prize. His previous book, *Eager: The Surprising, Secret Life of Beavers and Why They Matter*, won the PEN/E.O. Wilson Literary Science Writing Award. He lives in Colorado with his wife, Elise, and his dog, Kit—which is, of course, what you call a baby beaver.

ERIKA HAYASAKI is an independent journalist based in Southern California. She is a professor in the Literary Journalism Program at the University of California, Irvine.

FERRIS JABR is a contributing writer for *The New York Times Magazine* and the author of *Becoming Earth: How Our Planet Came to Life*. He has also written for *The New Yorker*, *The Atlantic*, *Harper's Magazine*, *National Geographic*, and *Scientific American*, among other publications. His work has received the support of fellowships from Yale, MIT, and UC Berkeley, as well as grants from the Pulitzer Center and the Whiting Foundation. He lives in Portland, Oregon, with his partner, Ryan, their dog, Jack, and more plants than they can count.

DHRUV KHULLAR, MD, MPP, is a physician and associate professor of health policy and economics at Weill Cornell Medical College. He is also a writer at *The New Yorker* covering medicine, health care, and politics. He serves as director of the Physicians Foundation Center for the Study of Physician Practice and Leadership, and as an Associate Director of the Cornell Health Policy Center.

SHARON LERNER is a reporter at ProPublica who writes about health and the environment. This story was a collaboration between ProPublica and *The New Yorker*.

MAX G. LEVY is a freelance journalist living in Los Angeles. He writes science stories for *Quanta Magazine*, *Wired*, TED-Ed, and elsewhere. He earned a PhD in chemical and biological engineering and is a co-founding writer of the independent science media site and newsletter *Sequencer*.

EMMA MARRIS is an environmental writer and the author of *Wild Souls: Freedom and Flourishing in the Non-Human World.* She lives with her husband and two children in Portland, Oregon.

JASON MAST is a general assignment reporter at *STAT* focused on the science behind new medicines and the systems and people that decide whether that science ever reaches patients. Before that, he worked as a biotech reporter at *Endpoints News.*

TOM MCALLISTER is the author of four books, most recently the novel *How to Be Safe* and the essay collection *It All Felt Impossible: 42 Years in 42 Essays.* His short fiction and essays have been published in *The New York Times, The Sun, Epoch, The Cincinnati Review,* and many other places. He teaches at Rutgers University–Camden and lives in New Jersey.

DAVID NAIMON is a writer and host of the podcast *Between the Covers.* His writing can be found in *Orion, AGNI, Tin House,* and *Boulevard,* among others. It has garnered a Pushcart Prize and been cited as notable in Best American Essays, Best American Mystery and Suspense, and Best American Travel Writing. He is also the author, with Ursula K. Le Guin, of *Ursula K. Le Guin: Conversations on Writing.*

EMILY RABOTEAU'S most recent book is *Lessons for Survival: Mothering Against "the Apocalypse."* She teaches creative writing at the City College of New York (CUNY) and lives with her family in the Bronx.

SCOTT SAYARE is a journalist whose work appears in *The New York Times Magazine, Harper's Magazine, The New Yorker, The Guardian*'s "The long read," and *New York,* among other publications. He lives in New York.

LEATH TONINO is the author of two essay collections, *The Animal One Thousand Miles Long: Seven Lengths of Vermont and Other Adventures* and *The West Will Swallow You.* A freelance writer, his prose and poetry appear in *Orion, The Sun, New England Review, Outside, Adventure Journal,* and elsewhere.

SARAH ZHANG is a staff writer at *The Atlantic,* where she covers science and health. Previously, she was a staff writer at *Wired.*

Other Notable Science and Nature Writing of 2024

Robin George Andrews
We May Be on the Brink of Finding the Real Planet Nine. *Scientific American*, December 17, 2024.

Tony Andrews
The Last Urchin Diver. *The Surfer's Journal*, October 1, 2024.

Christie Aschwanden
The Rise and Fall of Vitamin D. *Scientific American*, January 1, 2024.

Jacob Baynham
The Secret, Magical Life of Lithium. *Noema*, June 27, 2024.

Eric Boodman
"What's Your Pain Right Now?" Sickle Cell, Loss, and Survival in America. *STAT*, November 11, 2024.

Eric Boodman
This Federal Rule Didn't Stop Coercive Sterilization—But It Blocked Contraceptive Access.

Can It Be Fixed? *STAT*, June 18, 2024.

Marla Broadfoot
Families Under Attack. *Scientific American*, April 1, 2024.

Bethany Brookshire
Animals Are Our Neighbors in Cities and Suburbs, Not Pests. *Sierra*, March 14, 2024.

Jordana Cepelewicz
The Quest to Decode the Mandelbrot Set, Math's Famed Fractal. *Quanta Magazine*, January 26, 2024.

Andrew Chapman
Insatiable: A Life Without Eating. *Longreads*, April 18, 2024.

Kang-Chun Cheng
Sri Lanka's Catch-22. *Earth Island Journal*, July 2, 2024.

Siri Chilukuri
Industry Poisoned a Vibrant Black Neighborhood in Houston. Is a Buyout the Solution? *Grist*, March 6, 2024.

Christopher Cokinos
Fighting the Light. *Astronomy*,
June 4, 2024.

Eleanor Cummins
The Incredible Mystery of
NASA's Missing Moondust.
Popular Mechanics, January 25,
2024.

Tove Danovich
There Is No Show More Beautiful
Than This. *Emergence Magazine*,
October 8, 2024.

Catherine DeNardo
An A to Z of Hungry Killer
Whales. *Nautilus*, June 16, 2024.

Catrin Einhorn
Can We Invest in Solar Power
Without Harming Nature? *The New
York Times*, February 11, 2024.

Christian Elliott
The Secret Sex Lives of Deep,
Dark Corals. *Hakai*, November
27, 2024.

Delger Erdenesanaa
A Warning from a California
Marine Heat Wave. *The New York
Times*, December 1, 2024.

Monica Evans
The Coming Pollen Storms.
Noema, August 21, 2024.

Niall Firth
What We Leave Behind. *MIT
Technology Review*, September 1,
2024.

Amana Fontanella-Khan
Fibonacci Illuminati. *Pioneer
Works Broadcast*, January 16,
2024.

Devon Fredericksen
The Eider Keepers. *bioGraphic*,
March 8, 2024.

Elisa Gabbert
Fear as a Game. *The Believer*, July
11, 2024.

David Gessner
The Yellowstone to Yukon
Initiative Hopes to Give Wildlife
Room to Roam. *Sierra*, March 13,
2024.

Veronique Greenwood
The Once and Future Woods.
Nautilus, November 19, 2024.

Miles W. Griffis
Back from the Dead. *High
Country News*, June 1, 2024.

Rachel E. Gross
Women in Menopause Are
Getting Short Shrift. *The Atlantic*,
April 17, 2024.

Will Douglas Heaven
What Is AI? *MIT Technology
Review*, July 10, 2024.

Alexandra Horowitz
Best Inbred. *The New Yorker*, June
24, 2024.

Patrick House
The Lifelike Illusions of AI. *The
New Yorker*, March 19, 2024.

Sabrina Imbler
What Do We Owe Zoo Animals?
Defector, February 28, 2024.

Sabrina Imbler
How I Learned to Moth and
Embrace the Dark. *Defector*, July
18, 2024.

Jude Isabella
The Owls Who Came from Away.
Hakai, July 2, 2024.

Laureli Ivanoff
All Lady Seal-Hunting Crew.
High Country News, May 1,
2024.

Brooke Jarvis
Sea Change. *The New Yorker*,
August 26, 2024.

Casey Johnston
Collagen Sits on a Throne of
Lies. *Ask a Swole Woman*, January
14, 2024.

Sam Jones
An Ethical Way Forward for
Indigenous Microbiome
Research. *Nature*, September 2,
2024.

Katy Kelleher
High Mountains, Ancient Shells,
and the Wonder of Deep Time.
Nautilus, August 7, 2024.

Elizabeth Kolbert
When the Ice Melts. *The New
Yorker*, October 7, 2024.

Freda Kreier
The Enigmatic Earthquake
Hotspot in America's Heartland.
Undark, July 15, 2024.

Diana Kruzman
A Protected Place. *bioGraphic*,
June 11, 2024.

Diana Kruzman
Forests of the Future. *Earth Island
Journal*, July 16, 2024.

Jonathan Lambert
All Life on Earth Today
Descended from a Single Cell.
Meet LUCA. *Quanta Magazine*,
November 20, 2024.

Jamaal Lemon
Thinkin' It Can't Happen to
You, and Then It Do—How
Water Became a Casualty of
the Everglades' Seductive
Urbanization. *Good Beer Hunting*,
April 2, 2024.

Dave Levitan
The Time Paradox of Climate
Change. *Atmos*, April 8, 2024.

Andrew S. Lewis
The King Crab Kings. *Bloomberg
Businessweek*, May 29, 2024.

Shayla Love
How a Rare Disorder Makes
People See Monsters. *The New
Yorker*, August 1, 2024.

Shayla Love
The Longevity Hot Spots That
Weren't. *The New Republic*,
November 27, 2024.

Annie Lowrey
Why People Itch, and How to Stop
It. *The Atlantic*, October 23, 2024.

Annie Lowrey
The Truth About Organic Milk.
The Atlantic, April 12, 2024.

Adam Mahoney
Unsteady Ground. *High Country
News*, December 1, 2024.

Francesca Mari
Modern Warfare Is Breeding
Deadly Superbugs. Why? *The New
York Times Magazine*, January 26,
2024.

Kat McGowan
My Parents' Dementia Felt Like
the End of Joy. Then Came the
Robots. *Wired*, January 4, 2024.

Calli McMurray
A Scientific Fraud. An
Investigation. A Lab in Recovery.
The Transmitter, October 4, 2024.

Jackie Flynn Mogensen
Unnatural Selection. *Mother Jones*,
February 20, 2024.

Rudy Molinek
The Long, Strange History

of Teflon, the Indestructible Product Nothing Seems to Stick To. *Smithsonian*, August 20, 2024.

Sally Montgomery
A Freediver Finds Belonging Without Breath. *SAPIENS*, May 2, 2024.

Grey Moran
Fungi Are Helping Farmers Unlock the Secrets of Soil Carbon. *Civil Eats*, April 11, 2024.

Emi Nietfeld
The Parents Who Want Daughters—and Daughters Only. *Slate*, May 7, 2024.

Michelle Nijhuis
What Is the Opposite of Oil Drilling? *The New Yorker*, June 7, 2024.

Rachel Nuwer
Lifting the Veil on Near-Death Experiences. *Scientific American*, May 14, 2024.

Jeff Oloizia
Picturing an End to Alzheimer's. *Madison Magazine*, August 7, 2024.

Alisa Opar
Where the Not-So-Wild Things Roam. *Audubon*, March 26, 2024.

Stephen Ornes
Europa Ho! *Technology Review*, February 19, 2024.

Marissa Ortega-Welch
Untrammeled. *High Country News*, June 1, 2024.

Helen Ouyang
The Race to Reinvent CPR. *The New York Times Magazine*, March 27, 2024.

Lois Parshley
Taking Climate Killers to Court. *The Lever*, July 29, 2024.

Miranda Rake
The Case Against Lockdown Drills. *Romper*, November 12, 2024.

Fletcher Reveley
Advances in Mind-Decoding Technologies Raise Hopes (and Worries). *Undark*, January 3, 2024.

Matt Ribel
Disease Detectives. *Washingtonian*, May 7, 2024.

Yasemin Saplakoglu
How AI Revolutionized Protein Science, but Didn't End It. *Quanta Magazine*, June 26, 2024.

Yasemin Saplakoglu
How the Human Brain Contends with the Strangeness of Zero. *Quanta Magazine*, October 18, 2024.

Ari Schneider
Yellowstone. *Mountain Gazette 201*, May 15, 2024.

Kathryn Schulz
Starburst. *The New Yorker*, March 4, 2024.

Joanne Silberner
Quiet! Our Loud World Is Making Us Sick. *Scientific American*, April 16, 2024.

Richard Sima
A Mystery Illness Stole Their Kids' Personalities. These Moms Fought for Answers. *Washington Post*, May 12, 2024.

Ramin Skibba
It's a Bird, It's a Plane, It's Space Trash! *Rolling Stone*, July 6, 2024.

Maggie Slepian
  Danger on the Divide. *Longreads*,
  August 15, 2024.
Ashley Smart
  Haunting the Human Genome
  Project: A Question of Consent.
  Undark, July 9, 2024.
Jordan Michael Smith
  The Scientist Using Bugs to
  Help Solve Murders. *Smithsonian*,
  January 4, 2024.
Emily Sohn
  Maiken Nedergaard's Power
  of Disruption. *The Transmitter*,
  February 26, 2024.
James Somers
  Getting a Grip. *The New Yorker*,
  November 25, 2024.
Zack St. George
  The Comet Strike Theory Just
  Won't Die. *The New York Times
  Magazine*, March 5, 2024.
Abe Streep
  "It Feels Impossible to Stay":
  The U.S. Needs Wildland
  Firefighters More Than Ever,
  but the Federal Government Is
  Losing Them. *ProPublica*, March
  16, 2024.
Elizabeth Svoboda
  Intervention at an Early Age
  May Hold Off the Onset of
  Depression. *Scientific American*,
  January 1, 2024.
Kim Tingley
  Nature's "Swiss Army Knife":
  What Can We Learn From

Venom? *The New York Times
  Magazine*, November 13, 2024.
Jimmy Tobias
  The Spreading Deer Plague. *The
  Nation*, November 18, 2024.
Caroline Tracey
  Wilson's Phalarope to the
  Rescue. *High Country News*, July
  1, 2024.
John Travis
  The Vault Guy. *Science*, June 7,
  2024.
Boyce Upholt
  The Wonderful River of Oz. *The
  Bitter Southerner*, September 20,
  2024.
Linda Villarosa
  The Disturbing Truth About
  Hair Relaxers. *The New York Times
  Magazine*, June 13, 2024.
Katherine J. Wu
  How Long Should a Species Stay
  on Life Support? *The Atlantic*,
  March 15, 2024.
Katherine J. Wu
  The Koala Paradox. *The Atlantic*,
  June 17, 2024.
Joe Zadeh
  A Digital Twin Might Just Save
  Your Life. *Noema*, March 21, 2024.
Andrew Zaleski
  This Feather Could Save Your
  Life. *Washingtonian*, June 1, 2024.
Carl Zimmer
  AI Is Learning What It Means
  to Be Alive. *The New York Times*,
  March 10, 2024.

Explore the rest of the series

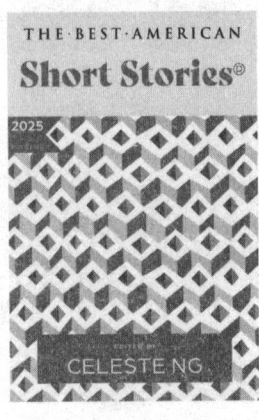

bestamericanseries.com